O9-AHV-913

U.S. CIVIL AIRCRAFT SERIES

VOLUME 1

This work is dedicated to the preservation and perpetuation of a fond memory for the men and the planes that made a future for our air industry. And, to help kindle a knowledge and awareness within us of our debt of gratitude we owe to the past. . . . .

First TAB printing 1993
Second TAB printing 1994
Published by TAB Books
TAB Books is a division of McGraw-Hill, Inc.

Printed in the United States of America.

Library of Congress Catalog Card Number 62-15967

U.S. CIVIL AIRCRAFT SERIES

VOLUME 1

(ATC 1 - ATC 100)

Joseph P. Juptner

TAB **AERO**

Division of McGraw-Hill, Inc.

Blue Ridge Summit, PA 17294-0850

ACKNOWLEDGEMENT

Any historian soon learns that in the process of digging for obscure facts and information, he must oftentimes rely on the help of numerous people; unselfish and generous people, many who were close to or actually participated in various incidents and events that make up this segment of history recorded here, and have been willing to give of their time and knowledge in behalf of this work. To these wonderful people I am greatly indebted, and I feel heart-felt gratitude. It is only fitting then, that I proclaim their identity in appreciation.

My thanks to Alfred V. Verville; Willis C. Brown; Tom Colby; W. U. Shaw; Chas. W. "Charlie" Meyers; Adolf Bechaud; S. J. "Steve" Wittman; Grover Loening; Albert Loening; Peter Altman; and Eddie Martin. To A. W. "Abe" French of Pan American World Airways; Ted McKee of Alexander Film Co.; Maxwell W. Balfour & Fred Tolley of Spartan Aircraft Co.; Don Black of Douglas Aircraft Co.; Gerald Deneau of Cessna Aircraft Co.; Herb Maxson of B. F. Goodrich Co.; F. J. Delear of Sikorsky Aircraft Div.; Ray Silvius of Western Air Lines; Monica Duston of Braniff International Airways; Lillian Keck of American Airlines; Jack Fraser of Boeing Airplane Co.; Elizabeth B. Brown of Institute of Aero. Sciences; John A. Cox & Harvey Lippincott of Pratt & Whitney Div.; Lewis L. Darling former editor of Western Aviation; Philip S. Hopkins and the staff at Smithsonian Institution, National Air Museum; and to Hartzell Industries; Fond Du Lac Chamber of Commerce; Fairchild Aircraft Div.; Defiance, O. Chamber of Commerce; Shell Oil Co.; Phillips Petroleum Co.; Texas State Chamber of Commerce; Temple, Texas Chamber of Commerce; Buhl Sons Co.; Eastern Air Lines; Richfield Oil Corp.; Parks College of Aero Tech.; Hawaiian Pineapple Co. (Dole); Beech Aircraft Corp.; Skyways Magazine; Texaco, Inc., Flight Magazine; Cleveland Pneumatic Tool Co.; Royal Canadian Air Force; Lockheed Aircraft Corp.; Northrop Corp.; United Air Lines; Convair Div.; Northwest Orient Air Lines; Berry Bros. Div.; Delta Air Lines; North American Aviation; Michigan Department of Aeronautics; American Aviation Historical Society; Eby Photo Service of Santa Ana, Calif.; and the following group of dedicated enthusiasts and historians; Stephen J. Hudek; Ken Molson; Roy Oberg; Leo J. Kohn; J. J. Sloan; Rex LaBreche; Frank Strnad; Jack McRae; Wm. T. Larkins; John W. Underwood; Marion Havelaar; Roger Besecker; Gordon Williams; Peter M. Bowers; and Gerald Balzer.

A detailed photo-credit list and a bibliography are included in the appendix.

FOREWORD

Just a short three or four decades ago, the sound of an airplane overhead as it flew serenely by, would cause heads for a mile around to turn upward; gazing up with shaded eyes to watch and to marvel at the contraption up there that was riding the wind so easily on it's flimsy wings. Or perhaps some would even watch with a little envy and inwardly yearn for a chance to lift themselves up into that vast sea of air and become unshackled, if only for a time, from a bond with the earth. Yes, the wonder and magic of flying was still very new to many, and the sight of an airplane flying overhead was indeed an occasion worthy of a few minutes pause.

Today, airplanes fly by only minutes apart, racing towards distant terminals which will be reached in a matter of hours ... and hardly a head is turned; the wonder of flying is no more new, it is no longer awe-inspiring. New contraptions continually vie for our notice & the airplane is more or less taken for granted. Surely, this was destined to happen, this is what hundreds upon hundreds of good men had dreamed for, strived for, and some even died for, but what is saddening to me at least, is the fact that so little is known and even less is remembered of the years of struggle that took place in the decades just gone by. A struggle in search for answers that often led to the point of heart-break and castastrophe in order to bring about this vast knowledge and comparitive maturity in the air that we now so casually enjoy.

In view of this, I humbly submit a plea that some obligation must be shown and a mark of credit recorded for the untold number of these pioneers in aviation, and the many creations they brought forth. A mark of credit for their boundless dreams that often knew no limit, and even an appreciation for their occasional folly; their every effort no matter even how small was another lesson learned, a point proven, and a valuable contribution added to the science of flight. Like another milestone on the path of evolution towards our modern airplane.

These following pages tho' they may sometimes sound like the nostalgic memories out of one man's past, are meant to be a factual story-telling account of little-known facets of these early formative years, and to exalt for a brief moment at least, the many efforts that contributed to the make-up of our early aircraft industry. I would say that the advent of the "Approved Type Certificate" for the manufacture of civil aircraft was more or less a charter for the beginning, it was at first a challenge that soon made way for opportunity and brought on an era of new dreams and new enthusiasm; it might actually be called the birth of the commercial aircraft industry.

A space of three years saw the aircraft industry emerge from a back-yard operation to one of the up-and-coming industries of the country. Of the many and varied offerings presented in the first three year period, the government agency, very tolerant and working feverishly, saw fit to certificate some 284 "types" of airplanes. This was far, far too many for the market available, to be sure, but many things were learned in the process, things that might have normally taken many more years.

With an intent to present this coverage in it's proper perspective and in some sort of sequence, we have chosen to list all of the airplane "types" that received the "Approved Type Certificate", because these were the ones that earned the stamp of approval and the blessings of the Dept. of Commerce to be built for the civil market. This then, for the most part, will present a cross-section of the scope of activity in the aircraft industry thru' all these years, and it's pattern of evolution towards the airplane of today.

Joseph P. Juptner

TABLE OF CONTENTS

1 Buhl-Verville "Airster" CA-3 (CA-3A), (Wright J4-J5). 9
2 Boeing "Model 40-A", (P & W Wasp 410-420). 14
3 Johnson "Twin 60", (2 Bristol "Cherub"). 17
4 Douglas "Model 0-2", (Liberty 12). 19
5 Douglas "Model M-2", (Liberty 12). 22
6 Douglas "Model M-4", (Liberty 12). 24
7 Alexander "Eaglerock", Combo-Wing, (Curtiss OX-5). 26
8 Alexander "Eaglerock", Long-Wing, (Curtiss OX-5). 30
9 Fokker "Universal", (Wright J4-J5). 33
10 Fairchild FC-2, (Wright J5). 36
11 Waco "Model 9", (Curtiss OX-5). 40
12 Buhl "Airsedan" CA-5, (Wright J5). 44
13 Waco "Model 10", (Curtiss OX-5). 46
14 Douglas "Transport" C-1, (Liberty 12). 49
15 Driggs "Dart" Model 2, (Anzani 35). 51
16 Stinson "Detroiter" SM-1, (Wright J5). 53
17 American Eagle A-1, (Curtiss OX-5). 56
18 Pitcairn "Mailwing" PA-5, (Wright J5). 58
19 Kreider-Reisner "Challenger" C-2 (KR-31), (Curtiss OX-5). 61
20 Fairchild FC-2W, (P & W Wasp 410-450). 64
21 OX-5 Swallow, (Curtiss OX-5). 66
22 Central States "Monocoupe", (Air Cat, Anzani 60). 69
23 Boeing "Flying Boat" B-1D, (Wright J5). 72
24 Stinson "Detroiter" SB-1, (Wright J4-J5). 74
25 Mahoney-Ryan "Brougham" B-1, (Wright J5). 77
26 Waco-Siemens "Model 125", (Ryan-Siemens 97-125). 81
27 Boeing "Model 40-B", (P & W Hornet). 83
28 Lincoln-Page LP-3, (Curtiss OX-5). 85
29 N.A.S. "Air King" Model 28, (Curtiss OX-5). 87
30 Travel Air "2000", (Curtiss OX-5). 89
31 Travel Air "3000", (Hisso A or E). 92
32 Travel Air "4000", (Wright J5). 94
33 Buhl "Airsedan" CA-5A, (Wright J5). 96
34 Loening "Cabin Amphibian", (P & W Wasp 410). 98
35 International F-17, (Curtiss OX-5). 101
36 Pheasant Model H-10, (Curtiss OX-5). 103
37 Travel Air "8000", (Fairchild-Caminez). 105
38 Travel Air "9000", (Ryan-Siemens 125). 107
39 Berliner "Parasol" CM-4, (Curtiss OX-5). 109
40 Curtiss "OX-5 Robin", (Curtiss OX-5). 111
41 Waco-Whirlwind ASO, (Wright J5). 113
42 Waco-Hisso DSO, (Hisso A or E). 115
43 Simplex "Red Arrow" K2S, (Kinner 75-90). 117
44 Simplex "Red Arrow" K2C, (Kinner 75-90). 119
45 Williams "Texas-Temple", (Wright J5). 121

46 Buhl "Sport Airsedan" CA-3C, (Wright J5). 123
47 Bellanca CH-200, (Wright J5). 125
48 Stinson "Junior" SM-2, (Warner "Scarab" 110). 128
49 Lockheed "Vega 1", (Wright J5). 130
50 Hisso-Swallow, (Hisso A or E). 133
51 Whirlwind-Swallow, (Wright J5). 135
52 Fokker "Super Universal", (P & W Wasp 420). 137
53 Arkansas "Command-Aire" 3C3, (Curtiss OX-5). 139
54 Boeing "Model 40-C", (P & W Wasp 410-450). 141
55 Stearman C3B, (Wright J5). 143
56 Fokker "Super Tri-Motor" F-10, (3 P & W Wasp). 146
57 Alexander "Eaglerock" A-1, (Wright J5). 149
58 Alexander "Eaglerock" A-2, (Curtiss OX-5). 151
59 Alexander "Eaglerock" A-3 & A-4, (Hisso A & E). 153
60 Sikorsky "Amphibion" S-38A, (2 P & W Wasp). 155
61 Fairchild FC-2W2, (P & W Wasp 420-450). 159
62 Stearman C3C, (Hisso A or E). 161
63 Curtiss "Challenger-Robin", (Curtiss "Challenger"). 163
64 Boeing "Flying Boat" B-1E, (P & W Wasp 420). 165
65 Cessna "Model AA", (10 cyl. Anzani 120). 167
66 Loening "Amphibian", (P & W Hornet 500). 170
67 Loening "Amphibian", (Wright "Cyclone" 500). 172
68 Curtiss-Robertson "Robin" B, (Curtiss OX-5). 174
69 Curtiss-Robertson "Robin" C, (Curtiss "Challegner"). 176
70 Mono Aircraft "Monocoupe" 70, (Velie 55). 178
71 Spartan C3-1, (Ryan-Siemens 125). 180
72 Cessna "Model AW", (Warner "Scarab" 110). 183
73 Spartan C3-2, (Walter 120-135). 186
74 Stinson "Detroiter" SM-1DA, (Wright J5). 188
75 Fairchild "FC-2 Challenger", (Curtiss "Challenger"). 190
76 Stinson "Detroiter" SM-1DB, (Wright J5). 191
77 Stinson "Detroiter" SM-1DC, (Wright J5). 193
78 Stinson "Detroiter" SM-1DD, (Wright J5). 195
79 Consolidated "Trusty" PT-1, (Wright-Hisso 180). 197
80 Consolidated "Husky" NY-1, (Wright J4-J5). 199
81 Consolidated "Husky" NY-2, (Wright J4-J5). 201
82 Consolidated "Courier" 0-17, (Wright J5). 203
83 Consolidated PT-3 & PT-3A, (Wright J5). 205
84 Consolidated "Husky Jr.", (Warner "Scarab" 110). 207
85 Hamilton "Metalplane" H-45, (P & W Wasp 410-450). 209
86 Laird-Commercial LC-B, (Wright J5). 212
87 Ford "Tri-Motor" 4-AT-B, (3 Wright J5). 215
88 Kreider-Reisner "Challenger" C-4A, (Comet 115-130). 219
89 Fairchild "Seventy One", (P & W Wasp 425-450). 221
90 Loening "Air Yacht" C2C, (Wright "Cyclone" 525). 224
91 Loening "Air Yacht" C2H, (P & W Hornet 525). 226
92 Pitcairn "Super Mailwing" PA-6, (Wright J5). 228
93 Lockheed "Vega 5", (P & W Wasp 425-450). 230
94 Hamilton "Metalplane" H-47, (P & W Hornet 525). 232
95 Mohawk "Pinto" MLV, (Velie 55). 234
96 Fokker "Deluxe Tri-Motor" F-10A, (3 P & W Wasp). 237
97 Kreider-Reisner "Challenger" C-3, (Warner "Scarab" 110). 239
98 Buhl "Senior Airsedan" CA-8A, (Wright "Cyclone" 525). 241
99 Buhl "Senior Airsedan" CA-8B, (P & W Hornet 525). 243
100 Travel Air "6000", (Wright J5). 245

A.T.C. #1
(3-27)
BUHL-VERVILLE "J4 AIRSTER", CA-3

Fig. 1. Buhl-Verville "J4 Airster" was awarded "Approved Type Certificate" No. 1 on March 29th of 1927; a memorable day that launched a new era in the annals of aircraft manufacture. Ship shown was flown to 2nd place in 1926 Ford Air Tour by Louis G. Meister.

Up through the years, the swift and phenomenal progress of aircraft development has had the ruthless tendency to obliterate memory and often push into near oblivion many airplane "types" that so generously helped to write the chapters of aviation's interesting history. Surely among those to suffer this undeserved fate was the Buhl-Verville "Airster". The "Airster" biplane, in it's brief history of just a few short years, had at least one major distinction that is not generally known and is more than likely nearly forgotten by now. In view of the circumstance, as it took shape, we might say that the "J4 Airster" had actually ushered in a brand new era to the annals of our aviation history.

The B/V "Airster" was the first of the civil or so-called "commercial airplanes" to earn and to receive the new "Approved Type Certificate". As time went on, this certificate became more commonly known to all as the "A.T.C.", and became highly coveted by those in airplane manufacture. These "certificates of airworthiness" were now being issued to aircraft manufacturers by the newly instituted "Aeronautics Branch" of the Dept. of Commerce. This move was designed to regulate the manufacture and help establish the "caliber" of the numerous aircraft that were in the process of being built and those new designs that were yet to be built for the "commercial" or civilian market. For a number of years up to this time, airplanes for civilian uses were being built according to the accumulated "know how" amassed by those active in the industry, and more or less on the honor-system, without any direct Federal supervision. But this system of things was found wanting in many instances and was deemed impractical in view of the terrific expansion forseen in the future for air travel and airplane manufacture. Therefore, under the Air Commerce Act of 1926, which was fathered by the Hon. Wm. P. McCracken, the "type certificate" became a regulatory measure and was made compulsory. An airplane "type" that had passed all the stringent requirements successfully, proved to one and all beyond reasonable doubt that it was completely airworthy and safe. The approved craft then received it's badge of honor, the "A.T.C.", and of course the blessings of the "Department". The "Number One" type certificate was awarded and issued to the Buhl-Verville "J4 Airster" with great honor and some fan-fare on the 29th of March in 1927. Since this time on, every airplane ever built to sell on the commercial or civilian market, has had to have and to faithfully conform to the provisions established in it's awarded "type certificate number".

The Buhl-Verville "Airster" as a type, had some previous history which began with the model CW-3. A handsome ship that was formally introduced and flown at Packard Field in Detroit by Jack Hunt in December of 1925. It was almost inevitable that the first of a "type" in these days, should be powered with the ever-popular and readily available Curtiss "OX-5", an 8 cylinder "vee type" engine of 90 h.p. This engine was also selected for the Buhl-Verville model CW-3, which started taking shape in March of 1925. Except for engine installation and a few minor differences, this prototype, shown here, was quite typical to the later type "J4 Airster". The CW-3 was a design actually well advanced for it's time, but safety was the prime consideration; among some of it's numerous interesting features were a rigid all-welded steel tube fuselage structure. It also featured a wide tread divided-axle landing gear and had folding wing panels, as shown here. The folding wing feature, for one reason and another, was thought to be quite important in these early days of aviation, and possibly it was to some extent. But as it turned out, it just never proved to be actually useful enough to warrant it's added expense; at least not in the U.S.A. where space was never at a premium. The performance and flight characteristics of the model CW-3 with the Curtiss OX-5 engine was really quite good for a ship of this type, and it seems likely that it would haved fared very well on the market; but the 1926 model of the "Airster" biplane was subsequently powered with the very popular, but still quite expensive, Wright J4 engine. The 9 cylinder Wright J4 was an air-cooled "radial type" engine of the "Whirlwind" series that was rated at 200 h.p. This added power produced a substantial increase in the airplane's utility and performance, but perhaps also narrowed down it's field of prospective buyers.

Later that year in 1926, a "J4 Airster" was flown in the Ford Air Tour by Louis Meister. Louie Meister, a former Air Service test-pilot of McCook Field, was now the company's chief pilot in charge of test and development. By a creditable endeavor, he finished a good strong second to the quite amazing Walter Beech who flew a new J4 powered "Travel Air" biplane. Finishing second to such a determined combination as "Beech in a Travel Air", was considered well enough indeed. Prior to it's certification, "Airster" production had been rather light, but through 1927 about 13 airplanes of this type were reported built. Two of these were the very first airplanes to be used by the new "Aeronautics Branch" of the Dept. of Commerce for their inspectors in field work about the country. This selection in itself, more or less endorsed it's worth and garnered a good amount of nation-wide prestige for this fine airplane. The first production model of the "J4 Airster", shown here in one of the accompanying illustrations, was bought by Henry B. DuPont of Wilmington, Delaware.

As it is pictured here in the various illustrations, we can see that the "Airster" was an average open cockpit biplane; having two open cockpits that seated a pilot and two passengers. This seating arrangement had fast become the standard layout for the light commercial or civilian type airplane of this period; and for some time to come. It was of a more or less conventional configuration

Fig. 2. The Buhl-Verville "OX-5 Airster", model CW-3, first flown in December of 1925 at Packard Field in Detroit.

Fig. 3. For space-saving storage, the wings of the "Airster" could be folded back to a width of 13.5 feet.

Fig. 4. This was the first production "Airster", and was bought by Henry B. DuPont of Wilmington, Delaware, for business and pleasure.

except that there was no stagger between upper and lower wing panels. The standard powerplant installation chosen for this model, the CA-3, was the reliable and thoroughly proven "J4 Whirlwind" engine of 200 h.p., but later in 1927, at least 3 of the "Airster" type were powered with the new "J5 Whirlwind" engine which was rated up to 220 h.p. A new slightly modified "J5 Airster", the model CA-3A, was flown by Nick B. Mamer, well known in the Pacific north-west, to 3rd place in the 1927 New York to Spokane Air Derby (Class A). Mamer was right on the heels of two hot-shot "Lairds" flown by Chas. "Speed" Holman and dapper E. E. Ballough, who finished one-two. The following year, a "J5 Airster" (CA-3A) was flown in the 1928 National Air Tour (which was formerly known as the Ford Air Tour) by Alger Graham and finished in 16th place amongst a stellar field of contestants which just about represented the cream of the industry's output.

Now, with all of the forgone in consideration, we must concede that the Buhl-Verville "Airster" had proven itself often enough to be a fine dependable airplane; it was never flashy nor did it have the compelling appeal that other airplanes of this type conveyed, but it did have a certain air of simple elegance about it, a sort of reserved and quiet charm.

Fig. 5. A ship of this type was flown by Nick B. Mamer to 3rd place in the 1927 Air Derby from New York to Spokane. Note modifications over earlier model.

Still and all, it found the going rather rough in nation-wide competition for sales and just couldn't manage to overcome the solidly entrenched popularity that airplanes like the "Waco" and "Travel Air" had built up and enjoyed. After a thorough check, it seems doubtful that any of this type are still in existence.

Alfred Victor Verville, who was once an engineer for Glenn Curtiss and designed and built his first airplane in 1915, shown here; is probably best known for his display of inventive genius in designing the "Verville-Packard" and the "Verville-Sperry" racing types that twice won the fabulous "Pulitzer Race" of the early "twenties". Alfred Verville was chief of design for the "Engineering Div." of the Air Service and was largely responsible for many innovations in airplane design; innovations which included such types as the VCP-1 and the PW-1, these became more or less the pattern for the famous and well-remembered "Boeing" and "Curtiss" pursuit series that followed later. Verville was of course the designer of the "Airster" series and it was first built in Detroit, Mich. by the Buhl-Verville Aircraft Co., which was organized in February of 1925. Detroit, already a bustling metropolis, was sincerely aspiring to become the "air capitol" of the U.S.A. The original B/V company was later reorganized into the Buhl Aircraft Co. and had moved to it's new enlarged plant in Marysville, Mich. Lawrence D. Buhl, a member of the "Buhl Hardware" family that was prominent in Detroit, was the firm's Pres. and Alfred V. Verville was the V.P. and Chief Engr. With sights set far ahead, Alfred Verville sold out his interests and left Buhl Aircraft later in 1927 to devote more time to the design and development of low-winged "Pursuit Ships" patterned after the successful "Verville-Sperry" R-3. These new designs were to feature the retractable landing gear, the low-winged configuration with an internally braced cantilever wing, and other outstanding features definitely new to the "pursuit type" airplane. But alas, all this was so far ahead of it's time that it was received rather skeptically and nothing ever came of it. Never one to rest on past laurels, Verville re-entered the commercial aircraft field in 1928 and designed and built the prototype of the handsome Verville "Air Coach" series; a discussion of this "type" will be in the chapter for A.T.C. #267.

Fig. 6. Airplanes of Verville design hark back to 1915. This was his first; lower photo on ice of Lake St. Clair; Verville in rear cockpit.

After the "Airster" type was more or less in decline, all efforts were turned to the developing and building of the Buhl "Airsedan" series under the capable supervision of Ettienne Dormoy; a fellow-engineer while with Alfred Verville in the "Engineering Div." of the Air Service, and certainly quite a versatile airplane designer in his own right. Dormoy is probably best known for his design of the "Dormoy Bathtub"; a rather unconventional ultra-lite airplane of 1924-25. The first of the "Airsedan" series was the model CA-5, see chapter for A.T.C. #12 in this volume. We might mention, Buhl did however, make one more try with an open cockpit biplane, a primary trainer type in an Army competition.

Listed below are specifications and performance data for the Buhl-Verville "J4 Airster"; span upper and lower 35 ft., chord both 60 in., wing area 300 sq ft., airfoil "Clark Y", length overall 25 ft., height 9 ft., width with wings folded 13.5 ft., empty wt. 1415, useful load 885, payload 440, gross wt. 2300 lb., max speed 125+, cruise 110, land 42, climb 950, ceiling 16,000 ft., gas cap. 40 gal. approx. range 440 miles. The following are "OX-5 Airster" specs and performance data; span, chord, wing area, airfoil, height, and width with wings folded were typical except for the following: length overall 25'6", empty wt. 1380, useful load 770, payload 350, gross wt. 2150 lb., max. speed 95+, cruise 80, land 40, climb 500, ceiling 10,000 ft., gas cap. 40 gal., oil 4 gal., approx. range 425 miles. The following figures are for the later type "Airster", model CA-3A, that was powered with the "J5 Whirlwind" engine; all specs were the same except for empty wt. of 1686, useful load 1383, payload 600+, and a gross wt. of 3069; these figures sound incredible but apparently were true. Performance figures were as follows; max. speed 125, cruise 108, land 52, climb 650, ceiling 12,000 ft., gas cap. in the fuselage and center-section tank 70 gal., approx. range 700 miles.

Also worth listing are some detailed features of the "Airster" type; the wing cellule was thoughtfully designed in such a manner that the upper and lower wing panels were interchangeable, as were the various components of the fixed and controllable tail surfaces. There was a generous amount of dihedral in the lower wing panels, but there was no interplane stagger. Bearing true to the time-honored practice of this period, the wing framework was built up of solid spruce spars and spruce and plywood built-up ribs, the panels were fabric covered. For space-saving storage, the wings could be easily folded back by two men in less than 5 minutes. The rugged fuselage framework was built up entirely of welded steel tubing, using no wire tie-rods for diagonal bracing; thus eliminating the fear of mis-alignment in service. The framework was lightly faired to shape with wood fairing strips and was also fabric covered. Cockpit and engine cowling was of sheet aluminum and a large entrance door was provided for the passenger convenience. All components were purposely built to a rather high safety factor, to stand up well under severe and rough usage that a ship of this type would normally encounter. This included the landing gear which was of the split-axle type employing "oleo struts" and rubber "compression discs" as shock absorbers; the landing gear was of the "tracking type", a Verville feature used by Navy carrier-based planes. The "tracking" feature only allowed a wheel movement up, and fore and aft, and therefore there was never any "tread change"; this naturally promoted better landings and eliminated sideward scuffing of the tires. The tailskid was steerable and wheel brakes were available; these were of the "Sauzzedde" internal-shoe type and were first used on the B/V "Airster". Proving entirely practical, these brakes later became the well known "Bendix brakes" a few years later.

Price at the factory for the "J4 Airster" was $9300, without folding wings, brakes, or dual controls; folding wings were $300 extra, brakes were $500 extra, and dual controls were $90 extra, metal propeller and inertia-type engine starter were standard equipment. The Wright "Whirlwind" J4 series engine developed 200 h.p. at 1800 r.p.m., it had an enviable record of performance that would make a story in itself. It weighed some 475 lbs. and cost over $5000, a cost which was usually prohibitive for a light commercial airplane of the type as used by the average fixed-base operator during this period.

A.T.C. #2
(7-27)
BOEING, MODEL 40-A

Fig. 7. The "Wasp" powered Boeing model 40-A for the first time made regular transcontinental air passenger service possible.

The year 1927 dawned with fervent activity and preparation the country over; an undercurrent of great promise was somehow felt and by the time the year was well under way, aviation was literally forced into the spotlight and was enjoying world-wide notice. This was due for the most part, to the number of ocean crossings and attempts from both of our shores, and the many other record-setting achievements that followed one another in rapid succession. Taking full advantage of it's good fortune, the industry followed up by working very hard to sell itself and it's services. So then by this time, interest in travel by means of the airplane was greatly stimulated and was generally on the increase, though the demand was still very erratic. To be ready on this demand was now the biggest problem. Existing equipment was fast becoming outmoded and was proving inadequate to serve in this capacity; it is then that those at Boeing realized that there was to be a definite need for an airplane that could be called upon to carry the few occasional passengers and their baggage, along with the normal everyday payload of air-mail and cargo. This then led to the logical development of the Boeing "Model 40-A", an airplane which soon demonstrated under service that it filled this need quite handily, at least for the time being.

As pictured here, the 40-A was a 3 place combination cabin and open cockpit biplane of rather large proportions, and of very rugged and dependable construction. The pilot was isolated from the cabin section in true "airmail fashion", in an open cockpit situated far aft in the fuselage for better "feel" and visibility, with a cabin section placed up forward that seated the two passengers. The passengers were fully enclosed and enjoyed a somewhat chummy comfort, the hatch-covered compartments for the mail and cargo were placed between the pilot's cockpit and the cabin section, and one just behind the engine fire-wall. The passenger's baggage, what little was allowed, was stowed in the lower section of this latter compartment.

The "Model 40-A" was of a configuration and construction that was to be fairly typical for all of the "40 series" that followed. In general description, it was a rather large "two bay" biplane with wings of equal span and the wing cellule employed no interplane stagger; a characteristic of design for a good many of the early biplanes. The powerplant used was the new 9 cylinder Pratt & Whitney "Wasp" air-cooled radial type engine of

Fig. 8. The "Liberty" powered Boeing Mailplane of 1925 (Model 40), from which the successful "40 series" was developed.

400-410 h.p., an engine that was introduced early in 1926 for use in Naval aircraft and was now gradually but firmly taking over in this power range since the venerable war-born "Liberty 12" was slowly dying out for use in commercial craft. The Model 40-A was the first commercial airplane to be especially designed for the "Wasp" engine. The development of the 40-A can largely be credited to Phil G. Johnson, who later became Boeing's head-man and remained at the helm for a good many years. The type certificate number was awarded to the 40-A in July of 1927 and it was the proud recipient of certificate "Number Two".

Twenty five of this model were built in 1927, the majority of them (24) for Boeing's use on it's own air-line system, the Boeing Air Transport, C.A.M. #18. Their San Francisco to Chicago run was formally inaugurated in July of 1927 and they were soon using a fleet of 28 airplanes to serve this route. Twenty four of these were the model 40-A and 4 of these were the newly developed tri-engined "Model 80", which operated between San Francisco and Salt Lake City. The "Model 40-A" was used into early 1928 and in a new development, the "Wasp" engine was replaced with the new and more powerful 9 cylinder Pratt & Whitney "Hornet" engine of 500 h.p.

Fig. 9. A portion of the fleet of Model 40-A airplanes to be used by Boeing Air Transport, are being here assembled out-doors at Sand Point.

This modification was redesignated the "Model 40-B" on another "type" number, see chapter for A.T.C. #27 in this volume.

Going back somewhat into the model's early history; the first of the "40 series" was actually the "Boeing Mailplane" of 1925, a ship that was designed for the competition held by the "Air Mail Dept." of the Post Office to procure an airplane design that would be more suitable for the transport of the air-mail than the obsolescent DH-4s that were then being used. The "Boeing Mailplane", shown here, was quite typical to the model 40-A in general, except that it was a single place open cockpit biplane and carried mail and cargo only. It was designed to carry a 1000 lb payload and was powered with the 12 cylinder "Liberty" engine of 400 h.p.; these were two requirements specified by the Post Office Dept. in it's design specifications. Unfortunately, the "Boeing 40" was not chosen by the P. O. Dept., the contract was let to Douglas Aircraft for some 50 of their "M series" biplanes, see chapter for A.T.C. #5 in this volume. In spite of this temporary set-back, the "Model 40 series", developed from the mailplane of 1925, later proved to be a useful and very popular type and had been progressively modified and improved in each successive model, the last of which was built sometime in 1931. The "Model 40-A" was manufactured in Seattle, Wash. by the Boeing Airplane Co. which was formed and started building airplanes under the guidance of W. E. "Bill" Boeing way back in the year of 1916, nearly a half-century of airplane manufacture to date.

Listed below are specifications and performance data for the "Wasp" powered Boeing "Model 40-A"; span upper and lower 44'2", chord both 79", wing area 547 sq ft., airfoil Boeing 103, length 33', height 12', wheel tread 88", empty wt. 3531, useful load 2469, payload 1600, gross wt. 6000 lb., max. speed 128, cruise 105, land 55, climb 770, ceiling 14,500 ft., gas cap. 140 gal., oil 12½ gal., approx. range 650 miles. The fuselage framework was built up of welded steel tubing, the rear portion was braced with steel tie-rods and was heavily faired to shape and fabric covered. The wing framework was built up of spruce spars and wood built-up ribs, also fabric covered. Of the tail-group, the horizontal stabilizer was built up somewhat like the wing panels with spruce spars and wood built-up ribs; all the other tail surfaces were built up of welded steel tubing, the tail-group was also fabric covered. All of the engine cowling, cabin doors, and compartment hatch covers were formed of "dural" sheet. The landing gear was of the typical Boeing cross-axle type and was built up of streamlined chrome-moly steel tubing, using Boeing "oleo-spring" shock absorbers and hand-operated individual wheel brakes.

Fig. 10. The Boeing 40-A topping the Ruby Mts. of Nevada on it's way across the continent.

A.T.C. #3
(3-27)
JOHNSON, "TWIN 60"

Fig. 11. Johnson "Twin 60", one of the most unusual light airplanes of this early period.

Other than the recollections of a few old-timers and some sketchy accounts in early periodicals, very little detailed information and historical data has been available on this rare and exceedingly unusual type. Born of an idea calculated to offer a little more reliability to light-plane performance than was usually the case, the "Twin 60" was introduced locally with very little fan-fare in the latter part of 1926. Tested successfully amid avid interest, it seemed to have a bright future placed before it.

Looking very much like a giant bomber in miniature, it was a 2 place open cockpit light twin-engined biplane of a much more different configuration than was normally used for a light 2 passenger airplane. After considerable study of powerplants available, it was easy to decide to power it with two Bristol "Cherub" engines; the gutty little "Cherubs" were a 2 cylinder opposed type air-cooled engine of 30-36 h.p. each and were of British manufacture. The Bristol "Cherub", for a time, enjoyed a good measure of popularity in the U.S.A. because of the lack of reliable American engines in this power range; most were little more than converted motorcycle engines which ran rather poorly and often failed suddenly without any warning. This lack of suitable power was indeed a sad state of affairs for a budding industry, because there were surely many indications present that there would have honestly been considerable and lasting interest in light airplanes at this time, had suitable engines been available.

On the Johnson "Twin Sixty", the engines were mounted backwards or in "pusher fashion" in strut braced nacelles, one on either side of the fuselage in between the wing panels; using reverse-pitch or "pusher props", of course. These propellers were metal "bent slabs" of a type made by Curtiss-Reed, although wood props were also used. The views shown here, picture the interesting and odd configuration; odd for a light-plane, that is. The fuselage was built into a sort of hull shape with the two open cockpits in the forward section, the pilot flew from the front cockpit enjoying an excellent and quite unrestricted view of all that was around him. A door was provided to the rear cockpit for the passenger's ease of entry and exit. Among some of it's other more prominent features were the gravity-feed fuel tanks that were mounted in the upper wing panels over each engine, each tank supplying it's respective engine directly below. The undercarriage was of the traditional straight-axle and spreader-bar type that had shock absorbers of rubber discs in compression. The tail-group had "balanced" twin rudders to provide good directional control; the use of twin rudders was also unusual for a light airplane but was a very familiar feature on large multi-engined airplanes for many years. The spacious and robust looking fuselage was built up of a steel framework, more than likely steel tubing; the wing panels were of the usual wood spar and wood rib construction, all was fabric covered.

The "Twin 60" was hailed as a gallant and visionary effort in design for a reliable light airplane, but due to circumstance of the times, it failed to make any lasting impression on those in the trade. The use of twin engines was certainly for added safety and increased reliability, but it seems very likely that one little "Cherub", all by itself, even "turning it's heart out", would have had quite a struggle to keep this airplane flyable at it's gross load. The type certificate number for the Johnson "Twin 60" was issued in March of 1927 and it was built by the airplane manufacturing division of the Johnson Airplane Supply Co. at Dayton, Ohio. E. A. Johnson was Pres. and Dave E. Dunlap was V.P. and Chief Engr. There is no recorded information handy on how many were actually built, only the evidence of a prototype. The bi-motored "Twin Sixty" was designed by Dave Dunlap who was a capable engineer and aircraft designer of the old school, serving as engineer for a good many of the old airplane companys of that day, and was later with Douglas Aircraft for over twenty years after joining them sometime in 1936.

Johnson Aircraft Div. long had been interested in getting connected with a good airplane design to build and to market; starting in 1923-24, they had more than a casual interest in the slick-looking "Hartzell" biplane models FC-1 and FC-2. These were built by the firm that was later of "Hartzell Propeller" fame. The FC-1 and the FC-2 were a 2 place open cockpit biplane of rather clean design that were flown by the affable Walter Lees (later of Packard "Diesel" fame), who was winning easily, many of the "OX-5 races" about the country. A creditable endeavor to be sure, but nothing ever came of it. The 3 little "Driggs-Johnson" ultra-light monoplanes of 1924-25-26 were Johnson built, they were designed by Ivan H. Driggs, also an engineer and aircraft designer of considerable note. Ivan Driggs left Johnson later in 1926 and formed his own firm to build the "Dart" and "Skylark" biplanes, both very fine little airplanes. For account of the "Dart 2", see chapter for A.T.C. #15 in this volume.

Listed below are specifications and performance data for the Bristol "Cherub" powered Johnson "Twin 60"; span upper and lower 28', chord both 45", wing area 194 sq ft., length overall 21' height 7'10", empty wt. 800, useful load 520, payload 190, gross wt. 1320 lb., max. speed 75-85, cruise 60, land 25-30, climb 550, ceiling 8,000 ft., gas cap. 25 gal. (12½ gal. in each tank), range 480 miles. With the strut bracing of the engine nacelles and the outer wing struts, the wing truss resembled the two-bay type but was actually not, in the true sense. All interplane struts were of round steel tubing and were balsa-wood faired to a streamlined section. The fabric covered tail-group was also built up of welded steel tubing. Originally, this airplane used a normal type tail-skid, but later was tried out with a tail-wheel. From scanning the records, it is safe to say that this was Johnson's last endeavor to manufacture airplanes, they fared much better as a supply-house.

Fig. 12. To demonstrate utility, the "Twin 60" was flown from the river-bank of the Miami River in Ohio.

A.T.C. #4
(6-27)
DOUGLAS, MODEL 0-2

Fig. 13. The "Liberty" powered Douglas 0-2 of 1926-27. This airplane took on varied forms in it's ten year history.

Quite often, as time passes by, a name in aircraft will eventually become synonomous with a service. So then, it is not too surprising that when most people hear mention of "Douglas" airplanes, in these later times, they will invariably think of the sleek, fast and high flying modern passenger transports that wing their way across our skies. Winging their way in all directions, and around the clock. This very easily described the "Douglas" airplanes of today and recent years, but actually Douglas has come up thru' the years and had it's start quite modestly in the years of "way back when". It was one of our earlier pioneers in the aircraft industry, even before it actually was much of an industry in the true sense of the word, and managed to build quite a good number of the so-called "stick and wire" airplanes of years gone by.

The Douglas Co., a mere handful of young enthusiasts, started out in Santa Monica, Calif. as the Davis-Douglas Co. and it's first airplane, called the "Cloudster", was flown on it's maiden flight in February of 1921 by Eric Springer. Only one of this model was built, but the basic design was so excellent and efficient that by 1922-23 Douglas was quite busy building torpedo launching planes for the U.S. Navy. The "Douglas" reputation now slowly becoming apparent, was more firmly entrenched and heralded world-wide by the 'round the world flight of the "Douglas World Cruisers" in 1924; a daring and historic adventure that was the first circumnavigation of the globe by airplane! Building airplanes to a standard that was always kept high, with an emphasis on design that assured utility and reliability, these features soon earned the name a service reputation that has always been among the best and held in high esteem, even to this very day.

The model "O-2", in discussion here and pictured in the accompanying illustrations, was a typical Douglas "type" of this period and shows it's evolution, to some extent, from the famous "World Cruisers" of 1924. In it's basic form, the O-2 was an Air corps "observation airplane", designed to perform the many duties peculiar to this branch of the corps. Seating two occupants in tandem, in open cockpits; the pilot was seated in the front cockpit and the observer-gunner was seated in the rear cockpit, operating either flexible-mount machine guns, or an aerial camera. The powerplant used for most of the versions in the "O-2 series", was the vener-

Fig. 14. The "Liberty" powered Douglas O-2 of an early type. The airplane that replaced the "DH-4" as the Army's standard observation-plane.

able "Liberty 12" engine of 400-435 h.p. This war-born engine saw many years of service already, but it was still going strong in military service and the U.S. government had a pretty large stock left on hand. A casual observation would mark the 0-2 as a rather large open cockpit, single bay biplane, with wings that were of equal span and having no stagger between panels; this had been and was yet to be a typical layout for the "Douglas" biplanes of this period. The 0-2, in it's various versions, had good habits and flight characteristics, so consequently it was always a favorite with the pilots of the "Corps". In service, it's performance was ideal and really quite good for a ship of this type.

The first of the "O-2 series" was designed in 1924 and was introduced as an Air Service "observation airplane" early in 1925, this prototype was designated the "XO-2". It was also powered with the 12 cylinder "vee type" Liberty engine and had been selected over some ten other designs submitted in the 1924 competition for a contract award of 50 airplanes. This was one of the largest contracts to be let at this time. Before long, it was generally felt throughout the service, and with some sadness here and there, that the coming of the Douglas model O-2 marked the passing of the beloved old "DeHaviland 4" as the standard observation airplane for the Air Service.

A type certificate number for the model O-2 was issued in June of 1927, but this was more or less a mere formality because none were ever built for commercial use, except in

Fig. 15. The Douglas O-2H, used extensively by National Guard units.

Fig. 16. Unusual view of the Douglas "Cloudster" of 1921.

it's modified version that was developed by Douglas as the "M series"; for a descriptive discussion of these, see chapters for A.T.C. #5 and #6 in this volume. As a basic design, the model O-2 went into numerous and extensive modifications, and altogether a total of some 507 airplanes of this series were built in a production period of about ten years. The O-2 series were manufactured at Santa Monica, Calif. by the Douglas Co., with the benevolent Donald Douglas as it's Pres. and Harry Wetzel as it's managing genius.

Listed below are specifications and performance data for one model of the "Liberty" powered Douglas O-2, there were many and varied models in the O-2 series so naturally the figures given hete will vary to some extent with the specifications and performance data of other models in this series, the figures given can be considered average; span upper and lower 39'8", chord both 68", wing area 414 sq ft., airfoil "Clark Y", length 29'7", height 10'3", empty wt. 2985, useful load 1870, payload 600, crew wt. 400, gross wt. 4885 lb., max. speed 129, cruise 105, land 55, climb 940, ceiling 15,500 ft., gas cap. 132 gal., range 650 miles. The fuselage framework was built up of welded chrome-moly steel tubing, faired to shape and fabric covered. The wing framework was built up of spruce spars and wood built-up ribs, also fabric covered. The fuel tanks were two in number and were mounted in the lower wing stubs which were built integral with the fuselage framework, a 10 gal. gravity-feed tank was placed in the upper right wing root to serve as a positive fuel supply in case of emergency. The robust landing gear was of the split-axle type and was built up of welded chrome-moly steel tubing, using "oleo" type shock absorbers, and it had a wheel tread of 102". The tail-skid was of the swivelling type for better ground operation on the turf surfaces used at all airports during this period; there was no such thing as a concrete runway! Armament and equipment carried was typical for an observation airplane as used by the Air Corps at this time.

Additional comment regarding the O-2 series: all were pretty much alike, with only slight changes here and there, until the model O-2H of 1927-28, this one was greatly revised and did not resemble the earlier models. The wings were of unequal span, with the upper wing span being the greatest, the wings now had interplane stagger and there was a completely re-designed tail group; cockpits were modified and the nose-section was topped off with a big square-looking flat radiator. Despite all these extensive changes, it still managed to look typically "Douglas".

Fig. 17. The "Douglas World Cruisers" of 1924, which made the first circumnavigation of the world by air.

A.T.C. #5
(6-27)
DOUGLAS, MODEL M-2

Fig. 18. The "Liberty" powered Douglas M-2.

The model M-2 was a hard-working Douglas type that was primarily designed and developed as a mail-carrier, and it was certainly instrumental in helping a pioneering air-line, the "Western Air Express", to establish a near-perfect record in maintaining it's scheduled air-mail and passenger flights during it's first year of service. This line's service was formally inaugurated on April 17th in 1926, being organized by Harris M. "Pop" Hanshue and under the direction of C. C. Moseley. The Western Air Express route, C.A.M. #4, was from Los Angeles to Salt Lake City, and in part, it followed the old "Mormon Trail" over the burning deserts and thru the treacherous mountain passes. This service was new in the west, and was truly a historical event in a new phase of aviation that was just beginning to take shape. Western Air Express started hopefully and quite modestly with 5 of these Douglas biplanes and soon was able to increase this fleet to 7. Now, after serving the western area continually throughout all these years to the present, it is one of America's most popular airline systems. Whether it was the first or not is a moot question, but it is rightfully considered as America's oldest successful air-line.

As pictured here, the M-2 was basically a single place open cockpit biplane with two covered hatches up forward that afforded 58.5 cu. ft. of carrying capacity for a maximum payload of over 1000 lb. Or, the two forward cockpits could be easily converted to carrying a passenger or two. True, it didn't offer the passengers any great comfort or convenience, the passengers often having to sit all bundled up in "flying clothes", with mail-bags at their feet; but then, air travelers were of adventurous spirit and were hardy

Fig. 19. Harris M. "Pop" Hanshue, Western's first president, saw his long, uphill struggle to establish the airline meet with success as he handed up a mail-sack for the first flight.

individuals in these early times! These adventurous souls were truly pioneers and helped to make aviation history, they certainly deserve honorable mention.

The configuration of the model M-2 was a typical Douglas layout. It was a fairly large single bay biplane with wings of equal span, and having no stagger between panels. The entire airplane was extremely rugged in nature, with plenty of built-in stamina and good performance. It was well liked by the pilots, and especially so in "dirty weather"; this was no doubt due to the M-2's predictable habits and flying characteristics which seemed to instill a feeling of trust and confidence when the going got particularly rough. Besides being in service for "W.A.E.", the M-2 was also in use by other lines that were coming along. A number were also used earlier by the airmail division of the Post Office Dept. whose many routes were now gradually being taken over by contract mail-carriers.

The first of the "Douglas Mailplanes" was the model "M-1" shown here, it was introduced late in 1925 and it was quite typical to the later built M-2 except that it had an under-slung "tunnel type" coolant radiator; showing influence from the early model XO-2 and O-2 as built for the Army Air Service. To simplify production and maintenance, the model M-2 mounted a flat nose-type radiator and had a profusely louvered engine cowling that could be "unbuttoned" in a hurry. Much of this was similiar to the later model of the O-2, from which the bulk of the M-series were developed. In 1925, Douglas won the competition sponsored by the Post Office Dept. for a suitable design to replace the time-worn "DeHaviland 4", one of which is shown here. Douglas received a contract award for 50 airplanes of the "M type" and

Fig. 20. The "Liberty" powered Douglas M-1, the first of the popular M-series.

Fig. 21. The time-honored "DeHaviland 4", a famous mail-plane of the early "twenties", which was operated by the U. S. Post Office Dept.

many were put into service with the "Air Mail Div."; slated to be used on various portions of the expanded transcontinental air-mail system. The type certificate number for the model M-2 was issued in June of 1927 and it was built by the Douglas Co. at Santa Monica, Calif. where they were justifiably proud of their famous slogan, "First Around the World". A slogan commemorating the U. S. Army's around the world flight in 1924 by "Air Service" pilots flying "Douglas World Cruisers".

Listed below are specifications and performance data for the "Liberty 12" powered Douglas model M-2; span upper and lower 39'8", chord both 68", wing area 411 sq.ft., airfoil "Clark Y", length 29'3", height 10'3", empty wt. 2885, useful load 1870, payload (normal) 900, gross wt. 4755 lb., max. speed 140, cruise 118, land 54, climb 950, ceiling 16,500 ft., gas cap. 130 gal., approx. range 650-700 miles. The fuselage framework was built up of welded chrome-moly steel tubing, faired to shape with spruce fairing strips and fabric covered. There were two cargo hatches forward of the pilot's cockpit and one small compartment underneath his seat for a total carrying capacity of 58.5 cu. ft. The wing framework was built up of solid spruce spars and spruce and plywood built-up ribs, also fabric covered. The fuel tanks were two in number and were mounted in the lower wing stubs, which were built integral with the fuselage; a 10 gal. gravity tank was placed in the upper right wing root as reserve. The robust landing gear used oleo-spring shock absorbers and it had a wheel tread of 102". The tail-skid swivelled for ease of taxiing on the turf surfaces used at all airports of this day. The cargo hatches were quickly convertible into open cockpits for seating one or two passengers.

A.T.C. #6
(6-27)
DOUGLAS, MODEL M-4

Fig. 22. The "Liberty" powered Douglas M-4 had all the "romance" of the DH-4 type, but it was much faster and more efficient.

The Douglas model M-4 was the last production version in the versatile and hard-working "M" series", and saw much service with numerous air-lines before it was finally retired from active use in the favor of more specialized equipment; nevertheless racking up an impressive record for efficient service with dependability. In it's basic form, the M-4 was quite typical to the three previous models in the M-series, except for slight changes and some additions that were found desirable, or necessary, from service experience gained by the millions of miles flown by the various M-type in all parts of the country. The M-4 was basically a single place open cockpit biplane, but the mail-hatch just forward of the pilot's cockpit, had a removable panel and could be converted quickly to carry a passenger, if need be. Power was furnished by the venerable 12 cylinder "Liberty" engine which managed to ring up a surprisingly good record for reliability in it's approximate ten year history, despite the many stories to the contrary.

Performance and flight characteristics of the M-4 were pretty much the same as those previous models in this series. The type certificate number for this model was issued in June of 1927, and this version was built into 1928; the certificate was amended and re-issued in July of 1929, offering selection of a combination of wing panels with greater or lesser wing areas, according to operating dictates for the service extended. This later version was still powered with the "Liberty" engine, which was becoming obsolescent for this type of use, but served quite well in a fast boat running rum from Canada!

The Douglas model "M-3", also shown here, was another model in the M-series that was very much alike in most respects, and saw a good deal of night-flying air-mail service with National Air Transport (N.A.T.). They had some 18 of these ships that were equipped for night-flying service on the New York to Chicago and Chicago to Dallas routes.

Note the early "Douglas" emblem on the fin of the M-3 and M-4 pictured here. These two models were also built by the Douglas Co. at Santa Monica, California. The well known J. H. "Dutch" Kindleberger was chief engineer for Douglas at this time; joining the firm sometime in 1925 and stayed on in that capacity until 1934 when he left Douglas to help form the General Aviation Mfg. Co., which was later to become "North American Aviation".

Listed below are specifications and per-

Fig. 23. The "Liberty" powered Douglas M-3. This view clearly shows the lack of interplane "stagger", a familiar feature in all early Douglas biplanes.

Fig. 24. A Douglas M-3 of the N.A.T. (National Air Transport) line loading mail at Chicago. Note man on left with sawed-off shotgun.

formance data for the "Liberty 12" powered Douglas model M-4; span upper and lower 39'8", chord both 68", wing area 411 sq. ft., airfoil "Clark Y", length 29'7", height 10'3"; empty wt. 3400, useful load 1455, payload 500+, gross wt. 4855 lb., max. speed 142, cruise 120, land 57, climb 900, ceiling 16,000 ft., gas cap. 130 gal., approx. range 700 miles. These figures above were typical for the night-flying version. For the day-flying version the following figures will apply; empty wt. 2910, useful load 2058, payload 1000+, gross wt. 4968, performance varied only slightly. For detailed discussion of construction details and other pertinent data, see chapter for A.T.C. #5 in this volume. The construction details were very much similar for all models in the M-series.

A model "M-4 Modified", sort of a last ditch stand for the "type", was powered by the 9 cylinder Pratt & Whitney "Hornet" engine of 525 h.p., it was built under a Grp. 2 approval numbered 2-45 which was issued in March of 1929. This "Hornet" powered combination, being otherwise typical, had a much better performance than the "Liberty" powered version; however, only this one airplane was built because this design had run it's course and was losing it's former appeal. A former attraction that was now being overshadowed by the increased performance and greater utility of more specialized and modernized airplanes that were being built in 1929.

The later M-4 type, built as per re-issued certificate of July 1929, had allowable wts. as follows; (with small wings) empty wt. 3405, useful load 1495, payload variable, gross wt. 4900 lb.; (with large wings) empty wt. 3580, useful load 2195, payload variable, gross wt. 5775 lb., extra fuel allowed. The next Douglas airplane to receive certification, following the standard M-4, was the Douglas "Transport" model C-1. For discussion of this type, see chapter for A.T.C. #14 in this volume.

Fig. 25. A veteran M-4 which saw many years of service with Western Air Express, (now Western Air Lines).

A.T.C. #7
(4-27)
ALEXANDER "EAGLEROCK" COMBO-WING

Fig. 26. 1928 Eaglerock Combo-Wing with OX-5 engine, a very familiar sight on pasture-airports about the country.

Late in 1924, the Alexander Film Co., a commercial "film shorts" producer headed by J. Don Alexander, had one of the first of the new "OX-5 Swallow" biplanes that they were using as a "company ship" in connection with their promotion and sales activities. This airplane proved to be such an asset to their operations that they promptly became avid boosters and supporters for the use of airplanes in business. Forseeing the possible future and taking the initiative, sometime along in 1925, they hurriedly announced publicly to the fact that they were now entering the business of building airplanes for commercial uses. Just shortly after, in July or August of 1925, they brought out their first airplane which they had named the "Eaglerock"; it is shown here in one of the illustrations. "Eaglerock" was indeed a very appropriate name for this airplane, considering that it was built among the eagles and the rocks in mile-high Denver.

This initial effort was a pleasant looking airplane; a 3 or 4 place open cockpit biplane, and quite naturally, it was powered with an 8 cylinder vee-type Curtiss "OX-5" engine of 90 h.p. This airplane was designed and developed by Daniel Noonan, and it was planned to incorporate all of the advanced ideas for civil aircraft that were known at this time. Among it's many novel and interesting features, were "hi-lift" wings of equal span that were connected to a normal center-section cabane, but using a steel tube "Warren" type truss for the interplane bracing, instead of the normally used steel wire cables. The wings could also be folded back in just a few minutes, and the unusual bracing truss kept the wing panels "in rig" while in the folded position. It had the traditional axle and

Fig. 27. The first Alexander "Eaglerock" of 1925, an interesting design by Daniel Noonan.

spreader-bar type of landing gear, but instead of the normal type tail-skid, it was to have a large steerable tail-wheel to make it roadable! This was to be a prime feature of this new airplane, but having insufficient and inefficient braking, it was sometimes unmanageable. According to one of their optimistic ads announcing it's introduction, "It flies like an eagle and travels the roads like an automobile."; surely a very commendable effort in this respect, but unfortunately, it didn't turn out that way. The fuselage framework was built up of steel tubing but it was not welded, they used special steel clamps to make the joints; the framework was then lightly faired to shape and fabric covered. This new "Eaglerock" fairly bristled with innovations and had a pretty fair performance too, but initial tests had soon proven that it would never measure up to expectations or promises extended; so this design was hurriedly dropped and a new improved "Eaglerock" was quickly developed.

The second "Eaglerock", also shown here, was announced in Nov. of 1925 and was introduced and first flown in January of 1926. This time they had come up with a pretty good one, not so pert and trim perhaps but basically a good sound design. They had dropped the idea of folding wings, and also the "roadable" feature; and decided to try a more conventional type. This early 1926 model of the "Eaglerock", as pictured here, was blessed with an abundance of wing area. It used the newly developed "Clark Y" airfoil section in a high aspect ratio which altogether provided a very efficient wing arrangement, giving exceptional high altitude performance with a low horsepower. It had a sturdy straight-axle type landing gear and a newly shaped rudder, a rudder with a large "balance horn" that was similar to that used on the "Travel Air" and many other types at this time. Figuratively speaking, this "Eaglerock" was of the "long wing" type, that is to say, that the lower wings had the greatest span. "Long Wing" was a term most often used by flying-folk to describe an airplane that had the greatest span in it's lower wings, but Alexander used the word literally to describe it's extra wing area, of which it had plenty. Daniel Noonan was also the designer and engineer on this one, and from the summer of 1926 he had the able and eager assistance of Al W. Mooney; a talented young draftsman-engineer who later was to become Alexander's chief engineer, and an aircraft designer of considerable note, in his own right.

After considerable development and naturally some modifications, the 1927 model of the "Eaglerock" was introduced with fan-fare late in 1926. It was also of the familiar "long wing" type, but by now, a "short wing" version and a "combo-wing" version had also been developed. For a time, Alexander offered an optional arrangement whereby "Eaglerocks" could be ordered with a choice of 3 different wing areas and configurations. It could be ordered as a normal "Long Wing" type which is discussed in the next chapter (#8), or it could be ordered as a "combination wing" (Combo-Wing) which was a nórmal "Long-Wing" type with about 3 feet cut off of each

Fig. 28. The second Alexander "Eaglerock" was test-flown January 1926.

Fig. 29. Fuselage structure of the 1926 Eaglerock; that beautiful contraption up forward is the beloved Curtiss OX-5 engine!

lower wing; this making the upper wing having the greatest span. It could also be ordered as a "Short Wing" type, sometimes called the "clipped wing", which was a normal "Long-Wing" type with about 3 feet cut off of each wing tip, both upper and lower. This version is shown in the next chapter. The "clipped-wing" was extremely rare and it was soon dropped in favor of the two standard models which were respectively known as the "Long-Wing" and the "Combo-Wing". The Combo-Wing came to be known more or less as the short wing "Eaglerock", it still had ample wing area in spite of it's shorter lower wing and was very suitable for general purpose use in most parts of the country, possibly being most popular in the mid-west and in the east. The improved models for 1927 now sported a split-axle type landing gear of rugged proportions; the wing panels also had a considerable amount of interplane gap between them, which in itself remained a familiar characteristic of the "Eaglerock" biplanes throughout the series. Interplane stagger was fairly normal but the aspect ratio was quite high; the "Eaglerock's" wing cellule was well planned for highest efficiency.

The trim "Combo-Wing" of 1927 was built under this certificate number which was issued in April of 1927. In it's basic form, it was a 3 place open cockpit biplane, as shown here, and was powered with either a Curtiss "OX-5" engine of 90 h.p., or the Curtiss "OXX-6" engine of 100 h.p. Two OX-5 powered "Combo-Wings", flown by Leslie Miller and James Shelley Charles, came in 2nd and 3rd in the Class B division of the 1927 National Air Derby from New York to Spokane. Leslie Miller led the race a good part of the way, but lost out towards the end by 28 minutes to hard-flying Charlie Meyers in an OX-5 powered "Waco 10". J. S. Charles in the other Combo-Wing came in 3rd, about 1½ hours behind.

Genial Don Alexander and part of his staff had spent considerable time and effort in showing these 1927 "Eaglerocks" off around the country in an extensive demonstration tour; as a result they soon began to catch on. 'Twas not long after that this airplane became a very familiar sight on pasture-airports all over the country. The "Eaglerocks" were manufactured in ever-expanding quarters by the Alexander Aircraft Co., a division of Alexander Industries at Denver, Colorado. For a discussion of the "Long-Wing" version, see chapter for A.T.C. #8 in this volume.

Listed below are specifications and performance data for the OX-5 powered Alexander "Eaglerock" in the Combination-Wing (Combo-Wing) version; span upper 36', span lower 32', chord both 60", wing area 330 sq ft., airfoil

Fig. 30. 1928 Eaglerock "Combination-Wing" with OX-5 engine, plane shown was serial No. 467.

"Clark Y", length 24'11", height 9'11", empty wt. 1420, useful load 760, payload 365, gross wt. 2180, max. speed 100, cruise 85, land 38, climb 514, ceiling 10,200 ft., gas cap. 37 gal., oil 4 gal., range 395 miles. Price at factory was $2475, and went up to $2750, late in 1927. Wings wired for lights were $20 extra. The fuselage framework was built up of welded steel tubing, lightly faired to shape with wood fairing strips and fabric covered. The wing panels were built up of a framework consisting of solid spruce spars and wood built-up ribs, also fabric covered. The fabric covered tail-group was built up of welded steel tubing, the fin was ground adjustable to help counteract high r.p.m. torque, and the horizontal stabilizer was adjustable in flight. The complete structure of the "Eaglerock" had an average safety-factor of 7 plus, thus attesting to their dependability in rugged service. The "Eaglerocks" were the first OX-5 powered light commercial type airplanes to receive a type certificate, and for this an admiration must be shown; because seriously, many airplane manufacturers sort of shunned the Dept. of Commerce tests, looking upon them with mixed feelings and much misgivings.

A.T.C. #8
(4-27)
ALEXANDER "EAGLEROCK" LONG-WING

Fig. 31. A striking view of the Eaglerock "Long-Wing", blessed with an abundance of wing area.

The familiar "Long Wing" variety of the Alexander "Eaglerock" was a companion model in this series for 1927 and was basically similiar to the combination-wing version in all respects except that it had a good bit more wing area; it's lower wings had the greater span. In it's basic form, the Long-Wing was also a 3 place open cockpit biplane, as shown here, and it was powered with either the Curtiss "OX-5" engine of 90 h.p., or the Curtiss "OXX-6" engine of 100 h.p. The OX-5 version being the more popular because the OXX-6 was not nearly as cheap to buy and was a good deal harder to come by; only a small number of these engines were built. The "Eaglerock" was also offered with "Hisso" power as a feeder-route mail-carrier, but this installation was rarely if ever used.

The Long-Wing "Eaglerock" was well noted far and wide as a gentle and amiable airplane, and a good pilot could operate it out of a ridiculously small field with nonchalance and plenty of ease, even at higher altitudes. Memory still recalls how this trait was proven on numerous occasions by an "Eaglerock" owner-pilot who used to take-off crossways on the narrow runway strip, often with a full load. He explained away his reason for pulling this caper by saying that his ship didn't handle too well in the brisk cross-wind that sometimes blew almost at right angles to the strip. Nevertheless, most of the boys thought he was just a plain show-off.

The Long-Wing was quite popular in the west and especially so in the mountainous areas of the northwest and southwest, where the "Eaglerock's" abundance of wing area and lift really paid off in better high altitude performance. It's often been said that these "Eaglerocks" were rather gawky looking and somewhat homely in appearance, but once in the air they became a ship transformed and to all were beautiful! This then would seem to confirm the fact that many old-timers used to say; the homelier they are the better they fly, or words to that effect. From the many stories and reports heard around and about, we can assume that a few of these lovable old ships are still occasionally active and giving their pilots many hours of real old-time enjoyable flying.

The original "long wing" type was the second airplane to be built by Alexander, and was first introduced in January of 1926. Going into further development, it was modified to some extent and later that fall was introduced as the new "1927 model", which is shown here. Among some of it's more apparent improvements were a sturdy split-axle type landing gear and a re-designed tail group. Two of the OX-5 powered "Eaglerock" Long-Wings were participants in the 1926

Fig. 32. A Short-Wing "Eaglerock" (clipped-wing) powered with Curtiss OX-5 engine, a 1926 offering that remained a rare type.

Ford Air Tour and finished in 9th and 15th place. Cloyd Clevenger, Alexander's test pilot, flew a "Whirlwind" powered special "Long-Wing" in the 1927 Ford Air Tour, but was plagued with engine trouble and missed some of the required stops; he finished in 13th place amongst some tough competition. The 1927 "Eaglerock" sold well and became quite popular for student instruction, it was regularly used by over 100 "schools" around the country by the end of the year. The type certificate number for the Long-Wing model of the "Eaglerock" was issued in April of 1927, and accounting for both models being produced, 205 airplanes were built in that year. The "Long-Wing" version seemed to be slightly the more popular, these two models were continued into 1928. By that time, the new and improved "A series" models were introduced, see chapter for A.T.C. #58 in this volume.

The airplane flown by Cloyd Clevenger in

Fig. 33. Eaglerock Long-Wing fitted with "Edo" floats.

Fig. 34. Early 1927 model of the Eaglerock "Long-Wing", it's amiable characteristics won many followers.

Fig. 35. Eaglerock "Long-Wing" shown, was once owned by Nellie Wilhight who was first aviatrix in South Dakota.

the 1927 "Tour", was built as a special that was powered with a 9 cylinder Wright "Whirlwind" J5 engine of 220 h.p., this was really a rare piece and must have been a terrific airplane. Basically, it was typical to the standard version except for added wing ribs and rounded wing tips, some minor changes in the cabane strut layout of the center-section attachment and some changes in the landing gear. It seems that this would have become a very popular sport-model, but only one was built.

Previously, in the spring of 1926, the famous General Wm. "Billy" Mitchel, sometimes known as the "stormy petrel" of aviation, dropped in on the Alexander factory one day and was quite impressed with the way the "Eaglerocks" were being built; he ventured to predict a great future for so fine an enterprise. The capable Daniel Noonan was chief engineer and was now very ably assisted by the efforts of many good men; one of whom was Albert W. Mooney, a talented gent who did big things while at Alexander and who later came into some fame in his own right. A live-wire sales force was headed by James A. McInaney, who was pretty darned good at praising the various virtues of the "Eaglerock" line. The ever-expanding plant of the Alexander Aircraft Co. at Denver, Colorado was under the watchful and paternal guidance of J. Don Alexander. Cloyd Clevenger, also a man of considerable talent, was their chief pilot in charge of test and development. The "Eaglerock" was the first airplane to be sold on the time-payment plan here in the U.S.A., it was also the first "OX-5 powered" airplane to be certificated, and those entries on this list were indeed many.

The so-called "long wing configuration", where the lower wing had the greatest span, was a feature used on many other old favorites such as the first Curtiss "Carrier Pigedon", the Curtiss "Lark", the Mummert "Mercury", some of the "Woodson" biplanes, the "Air King", and many others. By 1928 the idea was slowly giving way to the equal-span and especially to the use of the longer upper wing; in some cases the lower wing was abbreviated to such an extent that the airplanes were becoming "sesquiplanes".

Listed below are specifications and performance data for the 1927 model of the OX-5 powered "Eaglerock" Long-Wing; span upper 36', span lower 38', chord both 60", wing area 360 sq. ft., airfoil Clark Y, length 24'11", height 9'11", empty wt. 1470, useful load 760, pay load 365, gross wt. 2230 lb., max. speed 92, cruise 80, land 35, climb 485, ceiling 11,500 ft., gas cap. 37 gal., oil 4 gal., range 380 miles. Price at factory was $2475 and went to $2750 late in 1927. Wings wired for lights were $20 extra. For construction details check the previous chapter (#7), both models were identical in this respect. The "Eaglerock" structure was built rugged, a structure designed to withstand and absorb the cruel punishment most always encountered in student pilot-training.

A.T.C. #9
(6-27)
FOKKER, "UNIVERSAL"

Fig. 36. Early Fokker "Universal" powered with Wright J4 engine.

Tho' long since removed from the American scene, "Fokker" is still a name that hardly needs any introduction. There's surely not very many of us, even today, that haven't heard about or that do not remember the exploits of the famous "Fokker" airplanes. For their's was a name of countless romantic achievements for many, many years, and they were airplanes that were a familiar and very well known sight practically the world over.

The "Universal" was about the fifth type that A.H.G. "Tony" Fokker had introduced in the U.S.A. up to this time, and it was also the first of the "Fokker" type to be designed, built, and certificated in this country. It was also one of the very first commercial airplanes to be designed especially for the 9 cylinder Wright "Whirlwind" J4 engine.

Anthony Fokker, often called the "intrepid Dutchman", was indeed a great personality and somewhat of a show-man; his greatest ambition was to sell to the world the merits of the airplane, and especially the "Fokker" airplane!

The first of the "Universals" in production were powered with the Wright "Whirlwind" of the J4 series, a 9 cylinder air-cooled "radial engine" that was rated at 200 h.p. The first of this model was brought out in 1925, just shortly after the introduction of the famous "Fokker Tri-Motor", which made such a big hit in this country. The 1927 model of the "Universal" type, as pictured here, was a 5 to 7 place high wing cabin monoplane with the pilot seated in an open cockpit up forward of the wing's leading edge; 4 to 6 passengers were seated in the enclosed cabin section placed directly under the wing. The one-piece wing was of a lightly braced cantilever all-wood construction and was covered with a thin wood veneer. The fuselage framework was built up of welded steel tubing; both the wing framework and the welded steel tube fuselage were a method of aircraft construction that was reputed to be one of Fokker's many inventions during the W.W. I period.

Destined to work hard for it's keep, the "Universal" was designed of simple lines, and void of frills and fancy appointments; it was designed to work without fear of marring the face of her lean beauty. At ease in any combination, as shown here, the "Universal" could be operated on wheels, skis or floats, and being so versatile made them quite popular with the so-called "bush pilots" of this period. Especially in the north-country of the U.S.A., and in Canada and Alaska. They carried mail, freight or passengers, and were also used to good advantage in aerial-photography and various types of exploration work. Their short-field performance was very good, they afforded simplicity of operation with

Fig. 37. A J5 powered "Universal" on floats, modified by Western Canada Airways.

Fig. 38. The Fokker "Universal" on skis.

plenty of built-in stamina, and they could normally carry a 900 lb. payload with a range of 300 miles. When lightly loaded, as for aerial-photography, they could cruise for more than 7 hours. A number of the early pioneering air-lines such as Colonial, Pacific Air Transport, and others, used the "Universal" with great success on their scheduled routes.

The type certificate number for the Fokker "Universal" was issued in June of 1927, and records show that 23 of this model were built in that year. With the application of some slight modifications, this type was built on thru 1928. The newer "Universal", also shown here, was basically similar except for the following changes; they were now powered with the new "Whirlwind" of the "J5 series" which was rated at 220 h.p., the pilot's cockpit was still of the open type but was enlarged somewhat to seat two, side by side. The landing gear was typical but the shock-struts now had metal streamlined cuffs over the shock-cord windings, and a few other minor changes were added here and there. Changes that were found advisable thru some 3 years of service experience. Though the "J5 Universal" had substantial improvements over the earlier series, it was not greatly in demand. Possibly because the new "Super Universal" was coming out by now with it's more attractive qualities of greater payload and much better performance. For a later version of the "Standard Universal", see chapter for A.T.C. #164. The "Universal" was manufactured by the Atlantic Aircraft Corp., a division of the Fokker Aircraft Corp. of America, at Teterboro Airport, Hasbrouck Hts., New Jersey. Anthony H. G. Fokker was the chief engineer and Robert Noorduyn was assistant to the president.

Fokker "Universals" were most always so busy working that they hadn't much time for air-races and such, but Jack Frye did fly a "Universal" to 2nd place in an efficiency-race during the "Air Races" held at Spokane in 1927. Accounts of the Fokker "Super Universal" will be found in the chapter for A.T.C. #52 in this volume.

Fig. 39. 1928 Fokker "Universal" with Wright J5 engine was used on many early airlines.

Listed below are specifications and performance data for the "Whirlwind" powered Fokker "Universal"; span 47'9", chord at root 96", chord at tip 72", wing area 341 sq. ft., airfoil "Fokker", length 33'3", height 8', wheel tread 120", wts. as a landplane; empty wt. 2192, useful load 1500-1800, payload 900-1200, gross wt. 3692-4000 lbs., max. speed 118-110, cruise 105-100, land 45-52, climb 700-650, ceiling 12,000-11,500 ft., gas cap. 78 gal., range up to 7 hours. As a seaplane fitted with "Hamilton" duralumin floats; empty wt. 2653, useful load 1347, payload 750, gross wt. 4000 lb., max. speed 110, cruise 100, land 50, climb 650, ceiling 11,000 ft., all other specs and data applied. Price at factory; landplane $14,200., as a seaplane $16,650. Most of the construction details were previously described; wing was of spruce box-type spars, plywood ribs and stringers, and wood veneer covered. Wings were wired for lights. The fuselage and tail-group were of welded steeltubing and were fabric covered. It is interesting to note that all control cables, horns, pulleys, and so forth, on the "Universal" were exposed for ease of inspection and maintenance; this advantage far offsetting the gain that would be realized through more aerodynamic cleanness.

A.T.C. #10
(7-27)
FAIRCHILD, FC-2

Fig. 40. A 1928 Fairchild FC-2, powered with "J5 Whirlwind". This type was known as the "turtle-back" Fairchild.

By virtue of it's own efforts, Fairchild has always been and still is a trusted and respected name in the world of aviation, and it was also one of the early pioneers in the development of commercial aircraft. Way back in the turn of the "twenties", Sherman Fairchild was then primarily interested in the development of the aerial camera and it's uses in aerial photography. So, quite naturally he had occasion to contract for many different types of airplanes to carry out this work, and later on had numerous types in his own fleet of "camera ships". But Fairchild, eventually, like many other airplane users at this time, that were continually being forced to compromise between the performance or the utility of the types available. It was then he decided that the only answer to that problem was to build an airplane themselves, to suit their own requirements. The designs of their first airplane were first submitted to various local manufacturers, but most agreed that it would indeed be quite costly to build such an airplane as proposed by Sherman Fairchild. It was then Fairchild's decision to build it himself, securing the facilities available at the former Lawrence Sperry plant in Farmingdale, Long Island for just this purpose. This then did solve their problems for proper equipment but it also eventually forced them, willingly of course, into the airplane manufacturing business; an enterprise that has continually existed to this day and continues on as one of our most versatile and progressive.

Fairchild's first airplane, the model FC-1, was designed and built according to requirements decided upon through past service experience with other types of airplanes, and consequently when it appeared along about June of 1926, this is how it shaped up. It was a fairly large high wing monoplane with a sturdy semi-cantilever wing; this configuration was decided upon because it offered the least restricted view to the ground below and was deemed to be ideal and the most practical approach for many reasons. It had a fully enclosed cabin section, seating the pilot and one or two passengers quite comfortably; also providing protection from the elements for the cameras and the occupants, and not restricting the visibility in any way. The powerplant chosen for this first model was the 8 cylinder vee-type Curtiss "OX-5" engine of 90 h.p., and this was only natural for many "first models" of this period. The "OX-5" engine, being sold as war-surplus, was available so cheaply and easily. Among the many advanced features incorporated into the FC-1 were folding wings; primarily for space-saving storage. Late in 1926, after many hours of routine tests, the model FC-1, shown here, was flown by Richard "Dick" Depew, Fairchild's vice president for sales, in the Ford Air Tour of that year to test it's reliability under the severe conditions that would be encountered, and it made a very creditable showing. After the "tour", later on in 1926, the ship was redesigned and modified somewhat to take the power of a 9 cylinder Wright "Whirlwind" J4 series engine of 200 h.p.; it now became the model FC-1A which is also shown here. Naturally, it showed a substantial increase in performance and utility, and the more powerful engine now allowed a seating for five. Exhaustive tests and demonstrations carried on early in 1927 proved beyond doubt it's sound design and stimulated buyer's interest considerably, to such an extent in fact that it was some time before Fairchild could get around to building a few ships for their own use! With a new plant built and production soon to come off the line at one airplane per week, Gloria Swanson the well-known movie actress, christened the first of these and gave them a proper and fitting send-off. Among the listed optional engine installations offered, other than the J4 "Whirlwind", were the 6 cylinder Curtiss C-6 engine of 160 h.p., the A and E "Hisso" engines and the new 4 cylinder Fairchild-Caminez engine of 120-135 h.p., which was also a Fairchild development. See chapter for A.T.C. #37 in this volume for a more detailed discussion on the "Caminez" engine.

Before production actually got away to a good start the airplane was redesigned and modified again and became the FC-2, shown here in the accompanying illustrations. It was a slightly bulkier version that was now powered with the new 9 cylinder "Whirlwind" of the "J5 series", with the Curtiss C-6 engine slated as an optional installation. It was still a five place high wing cabin monoplane and was basically typical in most all respects; it being largely improved in detail and the capacity of the cabin was enlarged for a heavier and bulkier payload. The folding wing feature was retained and the ship could be readily converted to operate as a float seaplane or on skis. The folding wings, tho' of controversial value, were a good feature to have at times; the wings could easily be folded back in two minutes, the airplane thus displacing a total space of 14 feet wide by 31 feet long, allowing a great saving in storage space. But this feature also proved to be of inestimable value when a forced landing sometimes got one down into an extra small field, with no possibility whatever of taking off again from so small a field, the wings were simply folded back and the airplane was towed or taxiied to terrain more suitable for a safe take-off! This was more or less told in jest, but it actually was known to happen on occasion.

A few of the early FC-2s were put into Canadian service as float-seaplanes, one is pictured here, and their adaptability to the tough and peculiar requirements of "bush flying" was soon established. Strangely enough, the "Fairchild" monoplanes had begun to build up an envied reputation by now in the far away corners of many parts of the world but not hardly so in the U.S.A.; the

Fig. 41. The original Fairchild FC-1, powered with Curtiss OX-5 engine. Shown here as participant in 1926 Ford Air Tour.

Fig. 42. The FC-1 modified by the installation of a Wright J4 engine, became FC-1A.

Americans it seemed, had not yet fully realized the potential utility of carrying cargo by air, especially into remote sections not easily reached by any other mode of transportation.

Unheralded and quite by chance, an early FC-2 on "Floats", similiar to the one shown here, was flown by Cy Caldwell, an intrepid air-man and also a writer of some note; it was being ferried to a customer in the West Indies and became the first airplane to carry international air-mail from Key West, Florida to Havana, Cuba. Pan American Airways, just recently organized, had the contract and the sacks of mail but no way just then, to get them to Cuba. As Caldwell was going that way anyhow he offered to put the mail aboard and delivered it! Many such incidents were ample proof that the dependable "FC-2" was an airplane that never "played to the stands", so to speak, but went on about it's chores while often making history in a very casual manner.

Late in 1927, when Chas. A. Lindbergh embarked upon his nationwide good will tour in the famous "Spirit of St. Louis", he was accompanied by a party of 3 that flew along with him in a "Fairchild" FC-2, one of the early "pinch-back" type that belonged to the "Aeronautics Branch" of the Dept. of Commerce. The FC-2 was flown by Phil Love, an old time buddy of Lindbergh's from Army and air-mail days who was now a "Dept." inspector. Included in this party as Lindbergh's aide, was Donald Keyhoe. The tour covered over 20,000 miles throughout the 48 states in the 3 months of Aug., Sept., and Oct. Keeping to a strict schedule, the tour was practically trouble-free and "Lindy's" comment on the utility and performance of the accompanying Fairchild FC-2 was one of very high regard.

In January of 1928, Ruth Nichols, the celebrated and very capable woman pilot, flew as a co-pilot to Harry Rogers with M. K. Lee aboard on a route survey for a proposed air-line from New York to Miami, and made the 1250 mile flight non-stop in 12 hours. They made the flight in one of the new "turtle-back" FC-2s fitted with twin floats; loaded for the trip they took off with a useful load of something like 2400 lbs! These "FC-2's" were known to lift tremendous loads and they could actually carry 7 people on short hops and often did. To explain the phrase of "turtle-back", as used here, these were the improved FC-2 models of 1928 and among varied improvements they did away with the earlier "pinch-back" or "razor-back" fuselage that was 3-cornered in cross-section aft of the wing. This pinch-back configuration was apparently to lighten the structure and provide clearance for the wings when folded back. The

Fig. 43. A Fairchild FC-2 in Canadian service, an early example of this type, shown here fitted with Fairchild floats.

"turtle-back" FC-2 had a fuselage that was 4-cornered all the way back. The difference between the two is apparent in the various views shown. Ruth Nichols, by the way, was with Fairchild on sales promotion for a time. Another "J5 Whirlwind" powered FC-2 was flown in the 1928 National Air Tour by R. "Dick" Pears and finished in 12th place amongst a stellar field of 28 entries.

The approval for a type certificate number for the Fairchild FC-2 was issued in July of 1927 and 43 airplanes with the "Whirlwind" engine were built in that year; plus 4 ships with the 6 cylinder Curtiss C-6 engine, that were built especially for the Curtiss Flying Service. The next development was the FC-2W, see chapter for A.T.C. #20 in this volume. The J5 powered FC-2 was put into production on June 1st of 1927, from then to January 31st of 1928, a total of 56 airplanes were built and sold. It has often been said, and proven without a doubt, that the "Fairchilds" were a remarkable breed; they were gentle, dependable and obedient, never seeming to complain of the task at hand no matter what nor how difficult; that is about the best that one can say for an airplane!

Listed below are specifications and performance data for the "Whirlwind J5" powered Fairchild FC-2; span 44', chord 84", wing area 290 sq ft., airfoil Goettingen 387 Mod., length 31", height 9', wts. as landplane; empty wt. 2160, useful load 1440, payload 820, gross wt. 3600 lb. Weights as seaplane; empty wt. 2427, useful load 1573, payload 953, gross wt. 4000 lb., the seaplanes were fitted with "Fairchild" pontoons. Performance as landplane; max. speed 122, cruise 105, land 49, climb 565, ceiling 11,500 ft., range 700 miles. Performance as seaplane; max. speed 115, cruise 100, land 53, climb 420, ceiling 10,500 ft., gas cap. (both versions) 85 gal., range 650 miles. Some of these figures vary slightly with other published figures but it is reasonable to assume that all figures come within the normal tolerance allowed; manufacturers figures often tend to be optimistic. The fuselage framework was built up of welded steel tubing, lightly faired to shape and fabric covered. The wing framework was built up of spruce box-type spars and spruce and plywood built-up ribs, also fabric covered. The fabric covered tail-group was also built up of welded steel tubing, the fin was ground adjustable and the horizontal stabilizer was adjustable in flight. The seats were quickly removed for the hauling of cargo. For the FC-2 version that was powered with the 6 cylinder Curtiss "Challenger" engine, see chapter for A.T.C. #75 in this volume. In 1930 several of the existing "FC-2 type" were converted at the factory to take the power of the new "Whirlwind" J6-9-300, and also the new "Wasp Jr." of 300 h.p.; for a discussion on these, see chapters for A.T.C. #357 and #358, these were the models "51" and "51A".

A.T.C. #11
(7-27)
"WACO," MODEL 9

Fig. 44. Waco 9 submitted for flight and stress tests by Air Corps at McCook Field test-center.

It might be said that fortunate indeed are those that can remember associating with these amiable and completely lovable old "Waco" biplanes. The earliest of the "Waco" type actually dated back to the turn of the "twenties", or as many an old-timer will say, "back in the days of Buck Weaver". Only a few ships were built now and then and only a fair amount of success was enjoyed, of course, but along in April of 1925 when the first "Model 9" was introduced, the turning point had come; it was received and accepted with nothing but the greatest interest and enthusiasm. Surely not, that it was so revolutionary or unusual in any sense, but because it was a tidy and efficient looking airplane of sound design and great appeal, and it was fortunate to be brought out at a very opportune time. A time when a good many of the flying-folk, sparked by developments of that year, were just waiting for an honest-to-goodness airplane of this type with a reasonable price-tag to come along. Well planned, the "Waco 9" certainly had many fine qualities to ensure it's success and it pleased the men who flew it. Response to the "Nine" was heartwarming to the little group that conceived it, and 30 planes of this type were built and sold in the first 4 months of production.

Pictured here in various illustrations, is ample proof of statement that it was quite an attractive airplane with an altogether pleasing and well balanced configuration. A configuration that was coupled with a good all-round reliability and performance; a performance that was certainly among the very best of this period. Being true to the standard pattern that was evolving for an airplane of this type, the "9" was a 3 place open cockpit biplane and was powered with what is now almost reverently considered in aviation lore as a "wonder of wonders", the 8 cylinder vee-type Curtiss "OX-5" engine of 90 h.p. This early "Nine" of 1925, as shown here, had the popular "hi-lift" wings of equal span with large cable-operated plain ailerons on the upper wing panels; panels that were simply trussed in a single bay to center-section cabane struts that formed an inverted vee. The landing gear was of a sturdy straight-axle type, a feature that still hung on and was quite prevalent on most airplanes of this time. The divided or split-axle type, though already used on a number of other airplanes, had not yet proven itself to everyone's satisfaction. As the latest accepted practice at this time was to mount the engine radiator somewhere in the air-stream, as a "free air" type, "Waco" naturally complied and simply hung their's under the leading edge of the upper wing. This placement was often considered as a bad choice by many, for a number of reasons that should be obvious, but it remained a "Waco" trade-mark for many years and everybody just learned to put up with it. The rudder on the "Nine" was of the typical "balanced-horn" type that was so popular at this time and the engine cylinders were partially exposed through the engine cowling, allowing quick inspection and simple maintenance to the valve-gear and the plumbing, and assisting in cylinder cooling to some extent.

Two of these early OX-5 powered "Model 9's" were entrants in the first Ford Air Tour of 1925, they were flown by Eddie Knapp and Lloyd Yost and did very well in the unofficial scoring. Later in the year, many of the "Model 9's" were doing quite well in the numerous "OX-5 races" that were held about the country. One of many, Doug Davis who later became famous in the fabulous Travel Air "Mystery Ship" racing monoplane, sort of broke into air-racing with an OX-5 powered "Waco 9".

The 1926 model of the "Nine" was more or less typical except that later in the year there was an aileron change, a change to the "balanced-horn" type and a slight improvement was added to the landing gear set-up. Three "Waco 9's" were entered in the Ford Air Tour for 1926, this was the first year for official scoring and the "Wacos" finished 4th, 5th, and 6th. Of these three "Wacos", two were powered with the 6 cylinder Curtiss C-6 engine of 160 h.p., and one was powered with the "Hisso A" (Hispano-Suiza) engine of 150 h.p. The "Waco" biplanes were soon to become a top contender in these annual "air tours" and hung up an impressive string of victories throughout the following 3 or 4 years.

The latest version of the improved "Nine", shown here, was virtually unchanged. The "OX-5" engine was to be more or less the standard powerplant installation, but engines such as the 8 cylinder Curtiss "OXX-6" of 100 h.p., the 6 cylinder in-line Curtiss C-6 of 160 h.p., and the "Hispano" model A engine of 150 h.p. were of course optional. The type certificate number for the "Model 9" was issued in July of 1927 and 65 of this type were reported built in that year. A "Waco 9" was static-tested for safety-factor by the Air Corps in April of 1927 at the Army's McCook Field test-center and it showed up with an average safety-factor of 9 in certain critical attitudes, whereby only 6.5 was required for this class of commercial type airplanes; yet further proof of "Waco" built-in stamina and it's rugged character as it was demonstrated even before this, in actual service. This favorable test and the subsequent certification of the "Model 9", helped to offset to only some degree, the heart-felt tragedy that befell those at "Waco" with the passing of Elwood "Sam" Junkin. A talented, ambitious, and amiable man that was largely responsible for the designing of the various "Wacos" that were built up to and including the "Nine"; the affable "Sam" Junkin was sorely missed by all who knew him. After this temporary set back, everyone rallied around Clayton Bruckner, and "Waco" continued in force; the latest of the "Model 9's" were continued in limited production for part of 1927, they were being sold at a reduced price as a companion model to the new de-luxe "Waco Ten". For a descriptive account of the new "Model 10", see chapter for A.T.C. #13 in this volume.

The "Waco 9" was always proving itself versatile towards new uses, it could be operated as a float-seaplane with "Edo" floats and was one of the first "OX-5 powered" airplanes to do so with any amount of success. While on this vein, it seems appropriate to relate an amusing story once heard long ago during a session of hangar-flying; a story that told of a "Nine" on floats, no doubt similiar to the one pictured here, that was passenger-hopping at a small resort lake and the poor frustrated pilot had to continually caution his "ticket-hawker" not to line up "two big fat ones" for the same load. There's a limit to what "the old OX" will get off with, he'd say. Above a certain weight she just wouldn't "unglue the pontoons" from the water!

Fig. 45. A batch of 1925 "Waco 9" airplanes on factory field at Troy, Ohio.

Fig. 46. A "Waco 9" on Edo floats, powered by OX-5 engine of 90 h.p.

In it's approximate 3 year run, the "Waco Nine" had built up a very good reputation and had many enthusiastic boosters, but even at that it had to give way to the new "Model 10" because the "Ten" was even that much better! The standard "Model 9" first sold for $2500, at the factory, and then was reduced to $2225; later to $2125, and finally came down to the low figure of $2025. The "Waco 9" was manufactured by the Advance Aircraft Co. at Troy, Ohio.

Like a good many other airplane companies of this early period, "Waco" had it's misty beginning in the "turn of the twenties". It so happened that Geo. "Buck" Weaver had been barnstorming through Ohio in 1919 and was joined by Charlie Meyers, who had also been barnstorming in various parts of the country. Along the way, they were joined by two young fellows named Clayton Bruckner and Elwood Junkin. Aspirations common to all in the troupe led to the formation of a company, of sorts, to build airplanes in the town of Lorain, Ohio. This company was labelled the "Weaver Aircraft Co." and was, to some extent, under the guidance of "Buck" Weaver. Weaver was a fairly well known airman of this period, who for a time previous had been mechanic for the inimitable E. M. "Mattie" Laird and was also flying "Laird-Swallows" on exhibition. "Buck" Weaver was unfortunate to meet with severe illness and passed away about midyear in 1924. Always active and well-liked, he was sorely missed by many of the old flying fraternity. The catchy trade-name of "Waco" had been coined from the title letters of the "Weaver Aircraft Co." and by coincidence, "Buck" Weaver liked the sound of "Waco" in view of his happy associations at Waco, Texas while a flying instructor for the Air Service during the first World War period. This early enterprise formed in about 1920, actually had but very little significance in the make-up of airplane manufacturing history, but among old-timers there is always a saying, "I remember Waco clear back to the Buck Weaver days"!

A more fruitful beginning must be credited to the valiant and diligent efforts of Elwood "Sam" Junkin and Clayton J. Bruckner, two young enthusiasts who became avidly interested in aviation way back in 1914 and became steadfast partners. "Waco" as we came to know it, was hopefully but quite eagerly started in rather modest quarters at Troy, Ohio in 1923, largely by the efforts of Junkin and Bruckner, who already had gained considerable experience and could boast of a few successful seasons as airplane manufacturers. Their first "Waco" at Troy, Ohio was the OX-5 powered "Model 6", a 3 place

Fig. 47. The first "Waco", which was called the "Cootie", built in 1920.

"Jenny type" airplane of 30 foot wing span. It had a top speed of some 88 m.p.h., a good all-round performance. Soon to follow was the 3 place "Model 7" which was considerably improved over the "Six", and then came the unusual "Model 8". The "Model 8" was a cabin biplane that was powered with a "Liberty 6" (Hall-Scott) engine and could carry 8 people! Junkin and Bruckner were both fine pilots and did their own testing and demonstrating, and often resorting to week-end passenger hopping for money to help meet the payroll!

It wasn't until the advent of the "Model 9" in 1925 that "Waco" airplanes came to be recognized on a more or less nation-wide level, their enthusiastic acceptance was indeed welcomed at the little "Waco" plant. Forty-seven of these "Nines" were built in 1925, some 164 were built in 1926 and 65 were reported built in 1927. In 1927, the "Nine" had to compete for sales with the new "Model 10"; the "Ten" was going great, it was selling exceptionally well and the "Nine" was slowly losing out, so it was finally discontinued from production. Beside the hundreds of small operators who used the "Waco 9" for a multitude of duties, it was also used on a few of the early air-lines. Clifford Ball operated an air-line from Pittsburgh to Cleveland, starting in April of 1927 used "Waco 9's" to start with. This line later became Pennsylvania-Central, and later became Capitol Airlines. Embry-Riddle also used "Waco 9's" on their passenger and express line from Louisville to Cleveland which was inaugurated in May of 1927. So it is evident that the beloved "Nine" certainly did it's share, and more, for the development of civil aviation. According to some reports from here and there, a few of these old-timers have been restored; they have been completely rebuilt to as-good-as-new and are still flying.

Listed below are specifications and performance data for the OX-5 powered "Waco 9"; span upper 31'7", span lower 29'4", chord both 62.5", wing area 290 sq ft., airfoil "Aeromarine", length 23'3", height 9'3", empty wt. 1320, useful load 780, payload 385, gross wt. 2100 lb., max.speed 92, cruise 79, land 35, climb 500, ceiling 12,000 ft., gas cap. 37 gal., oil 4 gal., approx. range 375 miles, price at factory ranged from $2500 when first introduced to $2025 when finally discontinued. The fuselage framework was built up of welded steel tubing, faired to shape with wood fairing strips and fabric covered. The wing panels were built up of spruce spars and wood built-up ribs, also fabric covered. Interplane struts were of steel tubing in streamlined section, and interplane bracing was of standard braided aircraft cable. All movable controls were cable operated.

A.T.C. #12
(9-27)
BUHL "AIRSEDAN", CA-5

Fig. 48. An early version of the handsome Buhl "Airsedan" model CA-5. It was powered with Wright "Whirlwind" J5 engine of 220 h.p.

This model as shown here in the various illustrations, was the first type in a long line of "Airsedans" (Standard, Sport, Junior, and Senior) that were built by the Buhl Aircraft Co. in a period of about 5 years. The stately looking "CA-5", in it's basic form, was a 5 place cabin biplane which seated the pilot and the 4 passengers comfortably and all together in an enclosed cabin section; it may seem a little odd, but this was still being considered as quite an innovation even at this time. Actually, the first known practical cabin airplane was introduced way back in 1912 by "Avro" (A. V. Roe) of England, but the general acceptance of the cabin type airplane was very slow in coming. For a long time the fallacy prevailed that the pilot had to be out in the open with the wind-stream in his face or he wouldn't be able to fly the ship with accuracy and safety! With the increase in flying, especially in commercial activity, there was a definite need for the weather-protection of a cabin type airplane and skeptical pilots soon learned that it was not impossible to fly safely and accurately while enclosed in the comfort and protection of a cabin, thus a time-worn fallacy was gradually overcome.

The Buhl "Airsedan", Model CA-5, was an airplane of buxom and quite pleasing proportions in spite of it's size, and it's performance was certainly very good for a ship of this type. Although it's lower wings were quite small in comparison to the uppers, it was still considered a true biplane type; it was on the later models of the "Airsedan" that they whittled away at the lower wings until they became true "sesqui-planes", which is a term used in describing the "monoplane and a half". A 9 cylinder Wright "Whirlwind" J5 engine of 220 h.p. powered this first "Airsedan" type, and it's type certificate number was issued in Sept. of 1927. According to reports given, 8 of this model were built in that year; this would include the one that was especially modified from the standard "Airsedan" for participation in the famous and somewhat ill-fated "Dole Derby". The "Dole" race was a dash across the Pacific Ocean from Oakland, Calif. to Hawaii. This Buhl entry, shown here, was piloted by "Auggie" Pedlar with V. P. Knope as navigator and Mildred Doran, a comely and daring young school-teacher, went along as helper-passenger. This airplane was appropriately named the "Miss Doran" in honor of their fair companion. After being long overdue, the airplane was considered lost on this flight and no word nor trace of it was ever found. W. E. "Bill" Erwin and his navigator Eichwaldt,

Fig. 49. The ill-fated "Miss Doran" that was lost at sea during "Dole Derby" flight to Hawaii.

in the "Swallow" monoplane, who failed to get their overloaded airplane off the ground for the race, were later lost at sea in search of the missing flyers. In retrospect, it is somewhat remarkable that the "Miss Doran" actually got off of the ground with a load of 3 people, their equipment, and 400 gals. of gasoline on 220 h.p. and 350 sq. ft. of wing area; this amounted to a loaded gross weight of over 5000 lbs! In a way, proof positive of the remarkable performance built into the efficient "Airsedan" design.

When Alfred Verville severed connections from Buhl Aircraft in 1927, Ettienne Dormoy was engaged to take over as chief of design and development. This was brought about upon Verville's suggestion, Verville and Dormoy were fellow-engineers in the "Engineering Div." of the Air Service during the early twenties. Ettienne Dormoy was brilliant, visionary, and quite versatile; he is probably best known for his ultra-lite "flying bathtub" built in 1924. Louis Meister, who was Buhl's chief pilot in charge of test and development, flew an "Airsedan" in the 1927 Ford Air Tour and finished strongly in tenth place. Other "Airsedans" were in service for various business firms, transporting officials on company business that was scattered about the country, A. C. Spark Plug Co. was one of these. The "Airsedans" were manufactured by the Buhl Aircraft Co. in it's new plant at Marysville, Mich.; Lawrence D. Buhl was Pres., A. H. Buhl was V.P., Herb Hughes was Gen. Mgr., and Louis G. Meister was chief pilot. Ettienne Dormoy was in charge of design and engineering, with John Easton and W. U. Shaw assisting him. The next development in the "Airsedan" series was the improved model CA-5A, see chapter for A.T.C. #33 in this volume.

Listed below are specifications and performance data for the "Whirlwind J5" powered Buhl "Airsedan" model CA-5; span upper 42', span lower 32'4", chord upper 72", chord lower 48", wing area 334 sq. ft. (350 sq. ft. for "Miss Doran"), airfoil Clark Y, length 27'8", height 8'10", empty wt. 2073, useful load 1627, payload 1000, gross wt. 3700 lb., max. speed 120, cruise 105, land 48, climb 700, ceiling 13,500 ft., gas cap. 70-90 gals., oil 5 gal. approx. range 650-850 miles. Price flyaway field at factory was $12,500. The fuselage framework was built up of welded chrome-moly steel tubing, faired to shape and fabric covered. The wing panels were built up of heavy spruce spars and spruce and plywood built-up ribs, the wings were also fabric covered. The lower wing stub was built integral with the fuselage framework and was also built up of welded chrome-moly steel tubing. The lower wing stubs were of heavy cross-section to absorb the stresses of the split-axle landing gear which was mounted to the stub and the lower fuselage. The landing gear on the "Airsedan" followed the "tracking" principle introduced earlier on the Buhl-Verville "Airster", this permitted flexing of the under-carriage under load but caused no tread changes which often deviated an airplane on it's course down the landing path and eliminated tire scuffing. Cabin windows on the "Airsedan" were extra large for good visibility and there was an entrance door on either side for the passengers, and also one for the pilot. The pilot's station was slightly elevated and offered excellent visibility for a cabin biplane type. The fabric covered tail-group was also built up of welded chrome-moly steel tubing, the fin was ground adjustable and the horizontal stabilizer was adjustable in flight. Both the rudder and the elevators used aerodynamic "balance horns" for light pressures and ease of control. An inerita-type engine starter, wheel brakes, and metal propeller were standard equipment. Edo "floats" were available for $2500 extra. The famous "Whirlwind" J5 engine weighed about 495 lbs. and cost $4950, a cost of $10 a pound; gas consumption at cruise r.p.m. was 11 gals. per hr.

A.T.C. #13
(10-27)
THE "WACO 10"

Fig. 50. A handsome view of the original "Waco 10", showing it's many improvements over the earlier "Nine".

Tucked away in the memory of literally thousands, are fond recollections of a happy association with the friendly and throughly lovable old "Waco 10" biplane, an airplane that should no doubt go down in aviation history as just about the most popular of this period. It's trim and simple beauty and pleasant character, together with it's gratifying performance with the OX-5 engine, made it extremely popular with the small operator and private-owner. Who used it for just about everything, from week-end passenger hopping, charter trips to out-of-way places, dual-instruction, rental of solo-time, and what have you. Even those that used their "Ten" largely for sport, often resorted to any one of the above mentioned schemes to help defray expenses. The "Waco 10" was a good airplane, in fact, it's popularity became so great that "the boys at Waco" were forced to turn out over 350 of the "Model Ten" with the Curtiss OX-5 engine, and some 15 or so with the "OXX-6" engine, in 1927 alone! Besides these, there was a small number built of the modified "Tens" with various other powerplant installations that will be found listed under the various certificate numbers covering these models. This, we must remember, was quite a sizeable production for these times. It would be reasonably safe to say that in the ensuing years, the "Waco" biplanes, in the various models, had developed into the favorite airplane of all time. If not in numbers sold, then at least in a following of unshaken faith and loyalty to the "type".

Very much like the previous "Model 9", the "Waco 10" was also a 3 place open cockpit biplane and was powered with the 8 cylinder Curtiss OX-5 engine of 90 h.p. Similiar to a basic extent, but greatly improved. Pictured here, is the early 1927 prototype model of the new deluxe "10", it was clearly the best looking airplane of this series. In the various other illustrations, we can see the "Ten's" obvious and many useful refinements over the previous "Model 9". Features such as roomy, well protected, and comfortable cockpits, an easy-entrance door provided for the passengers, adequate wind-shielding, and a streamlined head-rest for the pilot; a split-axle type long leg "oleo-spring" landing gear of fairly wide tread, which incidently was a "Waco" first, and a trouble-free "spring leaf" tail skid. The entire engine was now enclosed in a neat cowling that was quickly and easily removable for inspection or maintenance. The engine cowling, of the "breast form" type, as shown here on the prototype model, was soon changed to a less complex type that was simpler and could be fabricated more easily. The change is shown here in the various other illustrations. The engine radiator

Fig. 51. The original "Waco Ten", being flown here on it's maiden flight by Charlie Meyers. Note "breast-form" cowling which was used only on this first airplane.

placement was the same as on the "Nine"; in the center-section bay, just under the leading edge of the upper wing. The "Ten" also had a redesigned tail-group that did away with the familiar "balance horn" of the rudder and the "horns" were omitted from the ailerons; the fin was ground adjustable to counteract high r.p.m. torque and the horizontal stabilizer was adjustable in flight. There were now four narrow-chord plain ailerons, one attached to each wing panel, with differential action and positive actuation. All of these features and a few more, all nicely blended into a trim and functional airplane that had remained in service long enough to prove it's worth many times over.

The "Waco Ten" was first introduced sometime in April of 1927; Charlie Meyers, a man of resourceful talent, was largely responsible for the design of the "Waco Ten" series, with the production and tooling problems worked out by Clayton Bruckner and "Bud" Schulenburg. Simplicity and ease of manufacture was the forte of the "Waco 10" design; speaking well for the men responsible for it's development. Charlie Meyers, briefly associated with "Waco" a few years back, returned to the fold in late 1926, shortly after the passing away of amiable "Sam" Junkin, and started design work on the "Waco Ten" series. By mid-summer of 1927 an accelerated production was well under way and a good number of "Wacos" were already flying all over the country-side. The amazing thing about the "Waco 10" was the way it was being accepted at the grass-roots level by the small operators; in less than a year's time there seemed to be at least one "Waco 10" at every airport!

The 1927 Air Derby, from New York to Spokane, Washington was the first cross-country dash that afforded the "Ten" to prove it's mettle. Flown by Charlie Meyers, with "Tom" B. Colby as passenger, an OX-5 powered "Waco 10" won first place in the Class B division of this grueling derby; covering the 2352 miles in 30 hrs. and 23 mins. of flying time. Meyers had a 28 minute lead at the finish after a nip and tuck race with an "Eaglerock" for most of the way. Another "Waco 10", flown by Eddie Knapp, came in 6th place, and another one flown by Jack Ashcraft came in 7th. Six more "Waco 10's" were also participants in this race but dropped out at various stages for different reasons. Altogether there were 32 entries in this derby, of which 25 actually got started and only nine airplanes completed the full route. Earlier, in June-July of that year, four "Waco 10 Sports" with "Whirlwind J5" engines made their presence felt in the 1927 Ford Air Tour by taking 5th, 7th 9th, and 12th places; see account of this in chapter for A.T.C. #41 in this volume. In the 1928 Air Derby, from New York to Los Angeles, "Tex" Rankin, a pilot of well known prowess, flew an OX-5 powered "Waco 10" to 5th place ahead of all other "OX-5 powered" entries. And so it went from year to year; "Wacos" didn't spend all their time winning races but certainly could if urged to do so.

The type certificate number for the OX-5 powered "Waco 10" was issued in October of 1927; later on as the "Model 90", and the "GXE", it was continually built into 1930 or so when the available supply of the OX-5 engine finally gave out! Considering the great popularity of this model, and the large number that were built, it's not really too surprising that a good number of these venerable old-timers are still flying in various parts of the country. The "Waco 10" was manufactured

Fig. 52. An excellent view showing the Curtiss OX-5 engine, as mounted in the "Waco 10"; note "oleo-spring" landing gear which was a "Waco" first.

by the Advance Aircraft Co. at Troy, Ohio. Clayton J. Bruckner was Pres., Charlie Meyers was test-pilot in charge of design-development, and "Bud" Schulenburg managed production facilities. Charlie Van Sicklen was in charge of sales.

Among the other models built by Advance Aircraft in 1927, there were the 15 airplanes mentioned previously that were powered with the 8 cylinder vee-type Curtiss "OXX-6" engine, 1 with the Model A "Hisso" (Hispano-Suiza) engine, 4 with the "Ryan-Siemens" (Siemens-Halske) engines, and 19 with the "J5 Whirlwind" engine. There were also a few modifieds that were powered with the "J4 Whirlwind" engine and one test model that was powered with the 4 cylinder Fairchild-Caminez engine. See chapters for A.T.C. #41 and #42 in this volume for accounts of the "J5-Waco" and the "Hisso-Waco". For the next development following the OX-5 powered "Waco Ten", see chapter for A.T.C. #26 in this volume.

Listed below are specifications and performance data for the "OX-5" powered "Waco 10"; span upper 30'7", span lower 29'5", chord both 62.5", wing area 288 sq. ft., airfoil "Aeromarine Mod.", dihedral both wings 1 deg., incidence both wings 2½ deg., positive stagger 14", interplane gap 4'11", length 23'6", height 9', empty wt. 1200, useful load 825, payload 354, gross wt. 2025 lb., max. speed 97, cruise 84, land 37, climb 520, ceiling 12,000 ft., gas cap. 37 gal., oil 4 gal., range 385 miles. Price at factory in 1927 was $2460, then to $2385 in 1928, and in 1929 the price had went up to $3145.

The fuselage framework was built up of welded chrome-moly steel tubing, faired to shape with wood fairing strips and fabric covered. The wing panels were built up of solid spruce spars and wood built-up ribs, also fabric covered. The fabric covered tail-group was also built up of welded chrome-moly steel tubing. The early "Waco Ten" was finished in two tones of Berryloid "Benz Gray" (almost a green), with silver wings. The fuselage and tail-group were a light shade of Benz Gray" and the metal cowls were several shades darker. Later models often had blue, green, or maroon fuselage, with silver wings. But of course, custom finishes were available. All wing and center-section struts were of streamlined chrome-moly steel tubing, and the interplane bracing wires were of steel in streamlined section. The fuselage tank of 37 gallon capacity was good for 5½ hours cruising under normal conditions, a 17½ gallon center-section tank was available, but not generally used. "Edo" pontoons were available at $1100 extra.

Fig. 53. Three "Waco Tens" frolicking over the country-side, "Waco Sport" in background.

A.T.C. #14
(10-27)
DOUGLAS "TRANSPORT", C-1

Fig. 54. A C-1C, 1927 model of the Douglas "Transport".

The versatile and hard-working "Transport" as built by Douglas back in these times, was quite a large single engined cabin biplane that was basically a development from the world-famous "Douglas World Cruiser" of 1924, both are shown here for comparison. The basic configuration and general appearance of the C-1 was typical and was unmistakeably a Douglas "type" of this period. In the vast depths of the fuselage was an enclosed cabin section which had ample seating for 6 to 8 passengers, or all of the seating could be quickly and easily removed for cargo hauling; the spacious cabin provided a useable space of about 160 cu. ft. for a maximum payload that ranged up to 2500 pounds. The crew of two, consisting of a pilot and a copilot-mechanic, were seated up forward ahead of the wings in an open cockpit just behind the engine section, with an access door at their backs leading to the cabin section, or vice-versa. The huge "Transport" was powered with the venerable "Liberty 12" engine of 400-420 h.p., the proposed commercial version of this type was basically similiar to the model C-1 as used by the Air Service (Air Corps) as a utility-transport. In it's military version, the C-1 was destined to perform such chores as personnel-transport, whether it be a full load of ten enlisted men or just one General, and it was called upon to do cargo hauling also. A cargo which often consisted of spare engines, propellers, and other airplane parts that were hastily flown to stranded or stricken airplanes. It was also called on to act as an air-borne ambulance on occasion, and was even used for experimenting in the use of "mass parachute drop" techniques. In other words, "utility transport" was a broad term and covered a multitude of duties.

In 1926 one of these model's was used as a "pathfinder plane" for transporting newsmen and for mapping out the itinerary of the "Ford Air Tour", a gruelling test of stamina for both planes and pilots that took place later that year. Some of us may also recall the role that the Douglas C-1 played as a "tanker" in the refueling of the renowned "Question Mark", the Air Corps' Fokker "Tri-Motor", model C-2, which set an endurance record by staying aloft for over 150 hours while flying up and down the coast over So. California in January of 1929.

The Douglas model C-1 transport was first introduced early in 1925 upon the request of the Army Air Service, and except for some

Fig. 55. The "Liberty" powered C-1 transport as it was introduced in 1925. Note resemblance to famous "Douglas World Cruiser".

minor changes and improvements, remained basically the same when it received a type certificate number that was issued in October of 1927. About 26 of this type in various models were built for Army service, but there is no record of the number built for commercial uses, if any at all. Although the C-1 "Transport" would have been well suited and had good possibilities as a profitable commercial-carrier, it remained best known as an Air Service (Air Corps) utility transport. It is indeed somewhat hard to believe that this lumbering transport was actually the forerunner to the world-famous "Douglas Commercial" or "DC" type that we knew of in more recent years! The model C-1 utility transport was built by the Douglas Co. of Santa Monica, California.

Listed below are specifications and performance data for the "Liberty" powered Douglas "Transport" model C-1; span upper and lower 60', chord both 84", wing area 800 sq. ft., airfoil Clark Y, length overall 36', height 14', empty wt. 3900, normal useful load 2400, normal payload 1260, normal gross wt. 6300 lb; gross wt. as 9 passenger 6660, gross wt. with maximum payload of 2500 lbs. is 7440 lb., max. speed 110-120, cruise 95-105, land 53-57, climb 550-645, ceiling 10,000-14,000 ft., gas cap. in 2 fuel tanks, one in each lower wing stub for a total of 125 gallons, a gravity-fed reserve tank in upper wing of 15 gal., oil cap. 12½ gal., approx. range 600 miles.

The fuselage framework was built up of welded chrome-moly steel tubing and steel tie-rod bracing, faired to shape with wood fairing strips and fabric covered. Aluminum cowling panels were used completely around the engine section and the fuselage, back to the front spar of the lower wing stub. There was a removable section in the floor of the cabin to facilitate the loading of articles that were too large for the cabin door-way, a spare "Liberty" engine could be loaded and carried in this manner. The baggage compartment was located aft of the cabin section. The wing framework was built up of spruce box-type spars and spruce built-up ribs, all panels were fabric covered. All interplane struts were of "dural" tubing of streamlined section and the landing gear structure was built up of chrome-moly steel tubing.

Fig. 56. The world-girdling "Douglas World Cruiser" of 1924. Many Douglas designs were developed from this basic configuration.

A.T.C. #15
(10-27)
DRIGGS "DART", MODEL 2

Fig. 57. 1927 Driggs "Dart" Model 2 with 3 cyl. Anzani engine.

The Driggs "Dart" Model 2 was a distinct departure from the configuration of the earlier "Dart", the Model 1, which was an ultra-light high wing monoplane. The "Model 2" shown here, was an odd but quite refreshing approach to the peculiar problems of light airplane design, and a very interesting little open cockpit biplane. This "Dart" had seating for two and it was powered with a 3 cylinder Anzani engine (French) of some 35 h.p. As many of the old-timers will probably remember, this engine was not the very best in power-plants, and certainly not the ultimate in reliability; but there wasn't much else to choose from in this power range at this time. The "Dart 2" was not really a small airplane in size, but most likely could still be classed as an "ultra-light"; in view of it's loaded gross weight of 820 lbs. and an empty weight of only 450 lbs. With two sub-normal sized people aboard and about 5 gallons of gasoline in the tank; it would put out with a pretty fair performance. With the little Anzani "hitting on all 3" and churning away at top speed, the "Dart 2" was good for about 85 m.p.h. and could climb out of a field at 420 feet in the first minute. With only one aboard the performance increased considerably, almost to the point of becoming vivacious!

The most outstanding feature first noticed on the "Model 2" was the rather unconventional placing of the two cockpits. One cockpit was ahead of, and one cockpit behind the center-section strut cabane; this was no doubt necessary to secure the proper fore and aft balance. Another interesting feature and also quite unusual, was the "Warren truss" type of interplane bracing; a bracing of steel tubing which formed sort of a W and used no cables to absorb the flying and landing stresses. The landing gear was of the fairly typical split-axle type, and the fabric covered fuselage framework was built up of welded steel tubing into an extremely light and rigid structure. The wing panels were built up of solid spruce spars and stamped out "dural" metal ribs, the finished panels were fabric covered.

It is quite interesting to note that all of Driggs' earlier efforts in airplane design were light monoplanes; the first Driggs type of any note was the little DJ-1 monoplane of 1924. The DJ-1 was entered in the Air Races held at Dayton, Ohio that year and won the Dayton News Trophy for light airplanes. The second version, built in 1925, was improved over the first type, but still quite typical. In 1926 Driggs built the first "Dart", which was also a modification of the original DJ-1 type. All of these versions were single place ultra-light parasol type monoplanes with closed-in sides to form a cabin enclosure, and used a one-piece full cantilever (internally braced) wing of very high aspect ratio. These were certainly graceful and beautiful little airplanes of very good performance, and proved their mettle in some mighty rough going encountered while participating in various cross-country events and closed-course races. All of this despite the

Fig. 58. Note seating arrangement and unusual wing truss of this "Dart" model 2.

unpredictable reliability of the so-called light airplane engines of these times. Ivan H. Driggs was the designer of these charming little monoplanes; the first two were built in the shops of the Johnson Airplane Supply Co. at Dayton, Ohio. The Johnson company (aircraft division) also built the "Canary" and later the "Twin 60", which is described in the chapter for A.T.C. #3 in this volume.

The Driggs Aircraft Co. was first organized at Dayton, Ohio and later at Lansing, Mich. in 1926 with Ivan Driggs as the V.P. and Chief Engr.; their first model produced was a modified version of the former DJ-1 type and was called the "Dart" Model 1. Their next effort was a 2 place high wing enclosed monoplane called the "Coupe", and it was the first airplane to mount the Detroit "Air Cat" engine (formerly Rickenbacker). Ironically enough, both of these types were highly praised as engineering advancements; but perhaps they were far ahead of the times, for they didn't sell. This no doubt prompted a change and the development of the "Dart 2", which was built shortly after. The "Model 2" received it's approval for a type certificate number in October of 1927. This little airplane was being offered for sale throughout most of 1928, but then the certificate was cancelled; the "Dart 2" had very unsatisfactory "spin" characteristics, and it is for this reason that this certificate was cancelled. This condition was later remedied in later Driggs "types".

Ivan Howard Driggs had somewhat of a genius as an airplane designer, especially with the light airplanes. He was a mere lad when he went to work for the old Burgess Co. and he designed and built his first airplane back in 1915 in Lansing, Mich. This was just after he graduated from Michigan State College. He was once an engineer with the Dayton-Wright Airplane Co. and also with Consolidated Aircraft Co., as assistant engineer to Col. V. E. Clark; prior to striking out on his own. Driggs was also the designer of the rugged little "Skylark" biplanes that were a later development from the "Dart" Model 2 type; for details on these, see discussion on A.T.C. #303.

Listed below are specifications and performance data for the "Anzani" powered Driggs "Dart" Model 2; span upper 28', span lower 22', chord upper 44", chord lower 24", area upper 100 sq. ft., area lower 40 sq. ft., total wing area 140 sq. ft., length 19'6", height 6'2", empty wt. 450, useful load 370, payload 110 lb. (or "one scrawny little passenger"), gross wt. 820 lbs., max. speed 85, cruise 75, land 35, climb 420, gas cap. 15 gals., range 350 miles. Price at the factory was $1750 to $1900.

Fig. 59. This "Dart 2" was still flying in 1939; the 5th airplane of this series.

Fig. 60. All of Driggs' earlier efforts were light monoplanes. here is the "Dart" model 1 of 1926.

A.T.C. #16
(11-27)
STINSON "DETROITER", SM-1

Fig. 61. The beginning of Braniff Airways, one Stinson "Detroiter" SM-1 and 3 employees! Paul Braniff pilot, first scheduled flight June 20, 1928.

The spring-board to fame and fortune for many and disaster for some, this "Stinson", model SM-1, was the first of the illustrious "Detroiter" monoplanes. All of the previous "Detroiters" that were built by Stinson Aircraft had been the model SB-1 cabin biplanes. The biplane version was popular to an extent, had sold quite well for about two years and had managed to pick up quite a little world-wide fame for the type, but the monoplane configuration with it's decided advantages was destined to replace it very shortly.

An "SM-1" monoplane of this early type, actually the very first one built, was flown by "Eddie" Stinson in the 1927 Ford Air Tour and he brought her in to a rousing first place win. There were 14 entries in the "tour" that year and Leonard Flo came in 6th with another "Detroiter", but it was of the SB-1 biplane type. The tour-winning monoplane was then modified for long-distance flying and became the well known "Pride of Detroit". The "Pride" was flown by Wm. S. "Billy" Brock and Edw. F. Schlee across the Atlantic Ocean in the course of their "2/3rds around the world" flight. This hazardous flight was ended in Tokyo, Japan on Sept. 14, 1927 more or less by popular demand. Their stalwart ship, shown here, was shipped home by boat to the relief of all, including the intrepid pilots who by now realized the folly of their undertaking at this particular time.

Paul Redfern, the courageous one, also used one of this early type on his ill-fated flight to So. America, about which many conflicting and controversial stories still exist. On Aug. 25 of 1927 Paul Redfern flew his heavily loaded "Detroiter", the "Port of Brunswick", out of Brunswick, Georgia on his way to Rio de Janiero. In due course he lost his way and finally crash-landed in the jungle thicket of British Guiana. Some ten years later, after 13 expeditions into the jungle wilds to solve the mystery of his disappearance, the remains of his airplane were found in the impenetrable jungle, but could not possibly have been brought out. There was a reliable eye-witness to the "falling of the great bird" however, thus possibly proving once and for all that he did crash and perish in the jungle.

The "American Girl", also shown here, was another well-known Stinson "Detroiter" monoplane; it was flown by Geo. Haldeman and the comely Ruth Elder across the Atlantic Ocean, or nearly so. They had to crash-land in the sea near the Azores Islands. Luckily, they were picked up by a nearby steamer and brought in to safety in Oct. of 1927. From these few incidents alone we can gather that

Fig. 62. Stinson "Detroiter" SM-1 "Eddie" Stinson's first monoplane which won the Ford Air Tour of 1927, then flew two-thirds around the world as the "Pride of Detroit".

Fig. 63. NX1384 at Roosevelt Field, L. I., N. Y. Oct. 10, 1927. Ruth Elder looking out of cockpit.

the "Detroiter" monoplanes were kept very busy making history and 'twas said that just about every one built was off on some sort of record flight!

The type certificate number for the Stinson SM-1, as shown here in the various illustrations, was issued in November of 1927. In it's basic form, the SM-1 "Detroiter" was a 6 place fully enclosed, strut braced, high wing semi-cantilever monoplane that was powered with the 9 cylinder Wright "Whirlwind J5" engine of 220 h.p. These few early models of the "Detroiter" monoplane had the same type of split-axle landing gear that was used on the SB-1 cabin biplane, but the later version of the SM-1 had a wide tread out-rigger type undercarriage, using "oleo" shock absorber struts. Thirty six of these "Detroiter" monoplanes were reported built in 1927, built on the average of one or two a week. A good many of these, possibly the bulk of them, were used for some sort of record-flight or attempt. 1927 was a very eventful year in the annals of aviation history and the "Detroiter" monoplanes contributed often and eagerly.

The SM-1 monoplanes were buxom, rather large and almost awkward looking but they delivered a good performance and handled surprisingly well in spite of their size. There was one of these that hauled Sunday passengers to "see the city from the air" out of a little two-by-four pasture airport on weekends, and we often wondered and even marvelled how this could be possible! But many stories heard since, would seem to indicate that a good "Stinson Pilot" was indeed a special breed of man and could really perform many wonders with this airplane.

Fig. 64. Early air-lines and progressive business houses were attracted to utility and performance offered by the Stinson "Detroiter".

Fig. 65. This SM-1 "Detroiter" was Eddie Stinson's personal airplane, with it he logged many miles in promotion trips about the country.

"Eddie" Stinson, just "plain-folks" with an ever-ready smile was one of our best known pioneering pilots, a true "early bird" and he was considered by most as the "master aviator". He had such a tremendous amount of flying time and experience, which actually dated back to 1912. Oddly enough, he was taught to fly by his sister, the well-known Katherine Stinson; Marjorie Stinson, his other sister, was also an accomplished pilot.

No "Stinsons" have been built now for well over a decade, but there are still plenty of them flying about the country in various models, some are over 30 years old by now! The Stinson monoplanes of this early period were built at the Northville, Mich. plant throughout 1927 and 1928, later moving to an enlarged plant with adjoining airfield in Wayne, Mich. For a descriptive discussion of later developments in the SM-1 series, see chapters for A.T.C. #74, #76, #77 and #78, all in this volume. The next development following the SM-1 was the SM-2 "Junior Detroiter", see chapter for A.T.C. #48 in this volume.

Fig. 66. The later version of the 1927 "Detroiter" now had "oleo" landing gear.

Listed below are specifications and performance data for the "J5 Whirlwind" powered Stinson "Detroiter" model SM-1; span 45'10", chord 84", wing area 292 sq. ft., airfoil "Stinson" (Mod. M-6), length 32', height 8'3", empty wt. 1970, useful load 1515, payload 805, gross wt. 3485 lb., max. speed 122, cruise 105, land 55, climb 750, ceiling 14,000 ft., gas cap. 90 gals., range 700 miles. Price at factory was $12,000 to $12,500. The fuselage framework was built up of welded chrome-moly steel tubing, heavily faired to shape and fabric covered. The wing framework was built up of spruce spars and wood built-up ribs, also fabric covered. There were two fuel tanks, one placed in each wing root flanking the fuselage. The fabric covered tail-group was also built up of welded steel tubing, the fin was ground adjustable and the horizontal stabilizer was adjustable in flight. Wings were wired for navigation lights, inertia type engine starter, metal propeller, and wheel brakes were standard equipment. Pontoons for operating off the water were $2500 extra. For descriptive discussion of the Stinson SB-1 cabin biplane, see chapter for A.T.C. #24 in this volume.

A.T.C. #17
(11-27)
"AMERICAN EAGLE", A-1

Fig. 67. The 1927-28 "American Eagle", powered with OX-5 engine. About to take-off with a load of joy-hoppers; this was a steady and sometimes lucrative source of expense money.

Hard on the heels of other types being introduced along about this time, and also quite anxious to get in the running, was the first of the "American Eagle" series which was test-flown on the 9th of April 1926. A number of flights in this prototype proved it quite satisfactory and well comparable to accepted standards quite familiar to the enthusiastic little group that made up the American Eagle Aircraft Co. at this time. Though rightfully jubilant over their new success, they were not striving to create any sort of aeronautical wonder, it was but their sole aim to build a good sound, practical airplane along normally accepted lines, that was to incorporate many of the latest features found desirable and even necessary through a number of years of past experience.

Faithfully following the general pattern set for light commercial type airplanes, it was a 3 place open cockpit single-bay biplane and though some differences here and there were easily discernable, it was still quite typical to the numerous other light commercial types that were being built at this time. The powerplant used in this "American Eagle" was the ever-popular Curtiss OX-5 engine of 90 h.p., as we think of it though, it's not very likely that the beloved "O by 5" was ever thought of as a powerplant. It's influence on one was of such a personal nature that it invoked a feeling far beyond the scope of any such mechanical contrivance as a "powerplant". Close association recalls that the "OX" was called many things ranging from "dear old sweetheart" to names and descriptions a lot less complimentary, depending on her disposition, and how she was "revving up".

In it's early form, the wings of the 1926 and the early 1927 "American Eagle" were of equal span, except for the over-hang of the "balanced-horn" ailerons that were on the upper wing; the rudder was also of the aerodynamically balanced-horn type which was typical of many other airplanes of this period such as the "Travel Air", "Waco 9", "Laird Commercial", and others. All movable control surfaces were cable actuated and the landing gear was of the typical straight-axle and spreader-bar type. The engine radiator was hung below the engine section in a manner that became known as the "free air type". The early 1927 "American Eagle" is shown here in good likeness in one of the illustrations.

The 1927-28 model of the "A-1" was introduced sometime later in 1927, and is also

Fig. 68. A 1928 "American Eagle" at the Ford Airport in Dearborn, Mich. Ford Airport was a popular stop-over for pilots in the Great Lakes area.

pictured here. It was very much the same, but was improved in many of it's details. As a major change, the "balance horns" were discarded from the ailerons in favor of the unbalanced or plain-type, that were now actuated by a push-pull tube and a slightly different rudder shape was adopted. The engine cowling was changed and simplified somewhat and now sported a burnished metal finish that suited the lines of the "American Eagle" very well. As we may recall, this feature of burnishing swirls on the cowling was made famous by that memorable airplane, the "Spirit of St. Louis". The engine radiator was still hung below the fuselage and the landing gear was now of the typical split-axle type, having one spool of wound rubber shock-cord to snub the loads. The cockpits were modified slightly and a streamlined head-rest for the pilot was added.

Normally, this model came powered with the OX-5 engine of 90 h.p., but numerous power installations were optional and available; engines such as the Curtiss OXX-6 of 100 h.p., the 150-180 h.p. "Hisso" engines, the 9 cylinder Ryan-Siemens of 125 h.p., the 10 cylinder Anzani engine of 120 h.p., and the new 7 cylinder "Quick" radial engine of 125 h.p., were offered. One "American Eagle" with the "Quick" radial engine was built for a sportsman-pilot who had a yen to tour the country by airplane. The type certificate number for the model A-1 was issued in November of 1927, and 92 were reported built through that year. Later in 1928, previewed as the 1929 model, the A-1 was modified still further but the only obvious changes made were the use of 4 narrow-chord ailerons, and a slightly modified rudder shape. This type, later as the "A-129", with continued occasional modification, was built through 1929.

In the early part of 1928, anyone that bought an "American Eagle" was most always offered a dealer-ship if it happened to be the first ship sold in this certain area, and this type of deal held true, more or less, for many other types of airplanes built during this period. Therefore, it was not too unusual for an operator with one lone airplane, nestled in a small "T Hangar", to display proudly a sign proclaiming him a "dealer"!

The likeable "American Eagle", though never hailed as a truly outstanding airplane, was a pretty good average for these times; it flew well and had no great vices and though it never enjoyed the popularity of the "Waco" or the "Travel Air", it was well liked by a great many. There are some of these old "Eagles" known to be in existence even yet, and one or two are known to be flying, possibly someday more will be rebuilt and will fly again. The model A-1 was built by the American Eagle Aircraft Corp. at Kansas City, Missouri, an enterprise that was deftly guided into being by E. E. Porterfield Jr. as the Pres., who in later years organized the Porterfield Aircraft Co. which built the popular "Porterfield" light monoplanes of the "thirties".

Listed below are specifications and performance data for the OX-5 powered "American Eagle" model A-1; span upper and lower 30', chord both 62.5", wing area 300 sq. ft., airfoil "Aeromarine 2A Mod.", length 24'1"; height 8'4", empty wt. 1227, useful load 814, payload 350, gross wt. 2041 lb., max. speed 99, cruise 85, land 35, climb 500, ceiling 10,000 ft., gas cap. 35-42 gal., oil 4 gal., range 385-425 miles. Late in 1927 the price at the factory was $2450, then $2515, in mid-1928 it was $2815 and later raised to $2985.

The fuselage framework was built up of welded chrome-moly steel tubing, faired to shape with wood fairing strips and fabric covered. The wing framework was built up of solid spruce spars and wood built-up ribs, also fabric covered. The already described landing gear was built up of welded steel tubing of streamlined section. The tail-skid was a steel tube with a hardened "shoe" on the end to offset rapid wear, it was also rubber shock-cord sprung. The standard color scheme of the "Eagles" was all silver with burnished "swirls" on the engine and cockpit cowling, prop spinner, and wheel fairing discs. The "American Eagle" really was a handsome airplane and looked very nice done up this way.

A.T.C. #18
(11-27)
PITCAIRN "MAILWING", PA-5

Fig. 69. The jaunty "Mailwing" was a handsome airplane and possessed terrific performance. Carried 500 lb. of air-mail or cargo.

The model PA-5 shown here in various views, was the first of the illustrious and very popular "Mailwing" series, and the fifth "type" to be brought out by Pitcairn Aircraft. In the year of 1925, Harold F. Pitcairn of Pittsburgh Plate Glass, who was above all a devout air-enthusiast since 1915, engaged the services of the brilliant young Agnew Larsen as a designer-engineer. Engaged for the purpose of designing, developing, and building a commercial-type airplane suitable to sell on the slowly awakening "civilian market". This merger of dreams and talents was an amiable and fruitful association that has surely left an indelible impression on the air industry; an association that was to last for a good number of years, in fact, until Pitcairn finally retired from all aircraft manufacturing activities in the late "thirties".

Pitcairn's first model was proudly introduced late in 1925 and was designated the "Fleetwing". The "Fleetwing", as shown here, was a fairly large open cockpit biplane of very pleasing lines and arrangement, it was powered with a 6 cylinder Curtiss C-6 in-line engine of 160 h.p.; it seated 5 people in 3 separate cockpits, all out in the airy open! A racy looking "sesqui-wing" called the "Arrow" was next in line; this one was primarily built for competition in the 1926 air-races. It was an open cockpit biplane that normally seated three, and could be powered with either the 8 cylinder OX-5 engine of 90 h.p. or the 6 cylinder C-6 engine of 160 h.p. To race in different engine classes, it had demountable engine assemblies and engines could be switched from one to the other, in less than 45 minutes! The "Sesqui-Arrow" went like the proverbial "bat out of hell" and was just about the only good competition that Chas. "Casey" Jones had against his very rapid C-6 powered "Clipped-Wing Oriole". Shortly thereafter and the next in line was the model PA-3; this one was called the "Orowing", and aptly enough too, because it used the Curtiss "Oriole" wings that were assembled around a Pitcairn fuselage, the framework of which was built up of welded steel tubing. In it's basic form, as shown here, the "Orowing" was also a 3 place open cock-

Fig. 70. Pitcairn's first airplane. The beautiful "Fleetwing" that carried 5 people on 160 h.p.

Fig. 71. Harold F. Pitcairn and the "Orowing" PA-3.

pit biplane and was powered with either the Curtiss OX-5, OXX-6, or C-6 engines. Early in 1927 a newly developed "Fleetwing", the "Fleetwing 2" was introduced, this was the model PA-4 and this time, in comparison to the previous "Fleetwing", was instead a 3 place open cockpit biplane and was powered with an OX-5 engine. It had a nose-type radiator and a sturdy axle-type landing gear; it was a handsome little airplane but due to circumstance, the possibilities of this ship were never fully developed, and only about 5 were built. The models PA-3 "Orowing" and the PA-4 "Fleetwing 2" were built and licensed under a Grp. 2 approval numbered 2-20 and 2-21, which were issued in November of 1927. This approval also covered all of these two models that were built prior to this date. The basic configuration of the "Fleetwing 2" was used for the prototype of the "Mailwing" series. This series were mainly developed for contemplated use on Pitcairn's eastern seaboard air-mail route that was formally launched on May 1st of 1928, and many years later was to become the vast net-work of Eastern Air Lines!

The model PA-5 was the first of the "Mailwing" series and was introduced about mid-1927. It was first built as a single place open cockpit biplane designed to carry air-mail or cargo in a hatch-covered compartment forward of the pilot's cockpit. The fuselage was deeply faired to form, and had ample room for a 500 pound payload which was handled quite well with the power of a Wright "Whirlwind" J5 engine of 220 h.p. Oddly enough, the prototype "Mailwing", shown here, still used the straight-axle landing gear, which was a carry-over from the "Fleetwing 2". Later production models were equipped with a split-axle "oleo" landing gear of the long leg outrigger type. Without question, the PA-5 was an extremely handsome airplane with plenty of heart and was delightfully lively; it's success as an air mail-carrier paved the way to national popularity for the models PA-6 and PA-7, two later versions of the "Mailwing" that were Pitcairn's best known. Twelve of the model PA-5 were reported built in 1927 and the majority of these were no doubt slated for use on the New York to Atlanta air-mail route C.A.M. #19, a route that was operated by a division of Pitcairn Aircraft. First flight on Pitcairn's line was flown by Eugene Brown on May 1st of 1928, a route of 792 miles serving 7 cities. C.A.M. #19, which was awarded in Feb. of 1928 soon had 16 PA-5 "Mailwings" in service, some of which were equipped with radio; these were among the very first to use radio in air transport. The versatility of the "Mailwing" soon became known and appreciated, other operators such as Clifford Ball (Pittsburgh to Cleveland), "T.A.T." in Texas, Colonial Air Transport, and others, were using them on newly awarded air-mail routes.

James Ray, Pitcairn's able test pilot, flew a PA-5 "Mailwing" to 11th place in the 1927 Ford Air Tour. James Ray also flew a PA-5 to 7th place (Class A) in the grueling 1927 Air Derby from New York to Spokane. At the air-races in Spokane, following the "derby", James Ray garnered a first place in a free-for-all race by turning the course at an average of better than 136 miles per hour, another PA-5 flown by A. M. Banks came in 5th. In the speed and efficiency race for the Aviation Town and Country Trophy, James Ray was awarded first

Fig. 72. Unusual shot of "Mailwing" on skis, in Canadian service.

place for speed (138+) and 3rd place for efficiency. This performance speaks well for the capabilities of the spirited "Mailwing".

A Pitcairn "Sport Mailwing", modified slightly from the standard "Mailwing", was exhibited for the first time at the All-American Aircraft Show held in Detroit in April of 1928. The "Sport Mailwing" was a 3 place open cockpit biplane and was leveled at the sportsman pilot; the passenger's cockpit could be covered with a demountable hatch, an appreciable increase in top speed was thus gained. The type certificate number for the model PA-5 was issued in Nov. of 1927 and it was built throughout most of 1928. For the next model in line, the PA-6 "Super Mailwing", see chapter for A.T.C. #92 in this volume.

Listed below are specifications and performance data for the "Whirlwind J5" powered Pitcairn "Mailwing" model PA-5; span upper 33', span lower 30', chord upper 54", chord lower 48", wing area 252 sq. ft., airfoil "Pitcairn", dihedral upper wing 1 deg., dihedral lower wing 4 deg., length 21'11", height 9', empty wt. 1612, useful load 1008, payload 500, gross wt. 2620 lb., max. speed 130, cruise 110, land 50, climb full load 1100, ceiling 18,000 ft., gas cap. 56 gal., range 600 miles. Price at factory was $10,000 later reduced to $9850. The specifications and performance figures for the "Sport Mailwing" were identical except for an increase in all-round performance, with a normal gross load of 2480 pounds. The fuselage framework was built up of welded chrome-moly steel tubing in a combination both round and square section, deeply faired to shape with wood fairing strips and fabric covered. The wing panel framework was built up of solid spruce spars and wood built-up ribs, also fabric covered. Torque tube operated airlerons were on the lower panels only, wing panels were wired for night-flying equipment. The fabric covered tail-group was also built up of welded chrome-moly steel tubing, the fin was ground adjustable and the horizontal stabilizer was adjustable in flight, rudder and elevators were cable operated. All interplane struts were of streamlined chrome-moly steel tubing, interplane bracing was of streamlined steel wire. Inertia type engine starter, metal prop, and wheel brakes, were available. The "Mailwing" in both the "standard" and "sport" series was manufactured by the Pitcairn Aircraft, Inc. at Bryn Athyn, Penn. with company headquarters in Philadelphia, Penn.

Fig. 73. The original "Mailwing" shown here at Ford Airport during 1927 Ford Air Tour; James Ray flew it to 11th place.

A.T.C. #19
(12-27)
K-R "CHALLENGER", C-2 (KR-31)

Fig. 74. The Kreider-Reisner "Challenger" biplane model KR-31, powered with OX-5, engine. This was one of the earliest examples.

It is logical to assume that Kreider and Reisner's happy and even profitable association with "Waco" airplanes in the years just previous, left them with some indelible memories and apparently this had influenced them a good deal towards investigating the attractive possibilities that were becoming apparent for the manufacture of commercial type airplanes, and spurred them on in the designing and developing of their own "Challenger" biplane. A casual glance would at first reveal that the "Challenger" type was basically similiar to the popular "Waco" biplane in many respects, but apart from a general resemblance, it certainly did have definite behavior characteristics and a personality that would be judged it's very own.

Consistent with general practice for an airplane of this type at this time, the "Challenger" was also a 3 place open cockpit biplane with a normal configuration. It was also powered with the Curtiss OX-5 engine. Among some of the more noticeable features of the C-2, as can be seen here in the various views, was the narrow, vertically mounted radiator of the "free air" type that was placed just below the upper wing, it's simple and sturdy landing gear that was calculated to stand up under the constant abuse encountered on the many "fields" that passed as airports in these days, and it's rather overly large and well rounded rudder. Performance-wise it was favorably comparable with the best of the "OX-5 powered" biplanes that were being built during this period. It flew quite well and handled nicely, it was well-mannered and had no bad habits to speak of and was always considered a good value for the money invested. The type certificate number for the "Challenger" C-2 (later known as the KR-31) was issued in December of 1927 and 9 of this type were reported built in that year. The "Challenger" biplanes were manufactured by the Kreider-Reisner Aircraft Co., starting out in very modest quarters at Hagerstown, Md.

Amos H. Kreider was a man of enthusiasm and considerable talent, and most of his earlier design efforts were with the ultra-light airplanes, during a period of time when this type of airplane was showing possibilities of becoming popular. The busy "K-R" shop built an ultra-light low wing monoplane in 1926, a ship that was an interesting design and quite successful. Also built in the K-R shop was a terrific little single place biplane that was designed and built by Charlie Meyers. Sometime later in 1927, Kreider and his partner L. E. Reisner, had mulled over plans to design and develop an airplane suitable for use by the numerous small-operators that were based at hundreds of scattered pasture-airports about the country. Being operators of a general flying-service themselves since 1925, they knew just what would be expected from a ship of this type; so they drew up plans and specifications for an airplane type they had in mind, and assumed it would have a fairly good chance of acceptance in this newly developed market. As it turned out, try as they might, they couldn't get anyone interested in it enough to build it, so they decided they might just as well go ahead and build it themselves. They were just aching to do it anyhow. This then marked the beginning of the popular "Challenger" line, a beginning that was very modest at first but slowly snow-balled into a rather impressive enterprise.

As the OX-5 powered C-2 was beginning to sell in encouraging amounts by 1928, K-R decided to develop new models. Modified somewhat from the model C-2, they developed and tested a version that was powered with the rather unusual 4 cylinder "Caminez" engine, a designation for this one has been rather elusive but this was more than likely the C-2. A C-3 version, modified still further, was also developed to take the 7 cylinder Warner "Scarab" engine, and then came the later-popular C-4A type that was powered with the 7 cylinder "Comet" engine, this model was soon put into regular production. The C-3 type remained quite rare and only a few were built, the "Caminez" version was only built in a prototype. For a descriptive discussion of the models C-3 and C-4A, see chapters for A.T.C. #88 and #97 in this volume.

1928 was a fairly good year for Kreider and Riesner, later on in the spring of 1929, K-R became a subsidiary of Fairchild Aircraft and continued to produce some real fine airplanes, among which were the notable "KR-34" series and the terrific little "KR-21" sport trainer. Partly as a demonstration to show the plane off and also as a shake-down test under the severe conditions that were sure to be encountered on a flight of this type, a "Challenger" model C-2, powered with an OX-5 engine, was flown in the Air Derby from New York to Los Angeles in the latter part of 1928 by Amos Kreider. He finished in 17th place after a grueling contest with most of the country's finest. There was a good number of these C-2 (KR-31) type built during a production period that lasted through the best part of 3 years and it is safe to assume that probably a good many of these are still around, somewhere. At least two of these were completely rebuilt and restored to like-new condition and are probably still flying. One of these rebuilt veterans is shown here.

Listed below are specifications and performance data for the OX-5 powered "Challenger" model C-2 (KR-31); span upper 30'1", span lower 29'2", chord both 63", wing area 296 sq. ft., airfoil "Aeromarine" Mod., length 23'9", height 9'4", empty wt. 1236, useful load 842, payload 360, gross wt. 2078 lb., max. speed 98, cruise 85, land 37, climb 548, ceiling 12,000 ft., gas cap. 33 gal., oil 4 gal., range 340 miles. Price at factory, less engine,

Fig. 75. A handsome view of the 1928 "Challenger" biplane, this ship was the 23rd off the line.

was $2240. The fuselage framework was built up of welded chrome-moly steel tubing, faired to shape with wood fairing strips and fabric covered. The wing panels were built up of solid spruce spars and wood built-up ribs, also fabric covered. The engine radiator was shuttered for temperature control. The OX-5 powered C-2 and the Caminez powered C-2 had 4 ailerons, all other models of the "Challenger" biplane had ailerons on the lower panels only. Built as special purpose airplanes, there was a KR-31 with a 5 cylinder Kinner engine and one with the Kimball "Bettle" engine. Later versions of the KR-31 were basically of the C-2 type.

Just in passing, and before we get on too far, here is an interesting note or two on the dearly beloved "OX-5 engine". Among other things, it was also refered to by some as the "O by 5". These simple and quite reliable old engines, even though they were built in 1917-18 for W. W. I trainers, played a very prominent part in the development of early commercial aviation. Primarily because they provided the manufacturers and operators with a suitable powerplant that was available at a very reasonable cost and readily available in great numbers. Some, many in fact, misjudged the merits of the "OX-5" and hastily predicted it's downfall into oblivion in the early "twenties", but they were surely surprised to learn that over 2000 of these engines were still flying as late as 1932! The OX-5, when in reasonably good shape, put out about 90 horsepower at 1450 r.p.m., it weighed 377 lbs. dry and the oil and water in it added some 80 lbs.; as a total weight with it's radiator and it's assorted plumbing, it weighed in at about 490 lbs. Full throttle gas consumption was about 9 gallons per hour, but at cruise r.p.m. a 35 gallon tank was often good for over 5 hours of flight or better than 400 miles. Through the years, there have been many discussions on the relative merits, and demerits, of the old OX-5 engine but it must be readily conceded that it was the best thing that ever happened to early commercial or civil aviation!

Fig. 76. 1929 KR-31 that was completely restored to flying condition, some 30 years after it's original manufacture.

A.T.C. #20
(12-27)
FAIRCHILD, FC-2W

Fig. 77. Fairchild FC-2W powered with "Wasp" engine.

The lean and brawny Fairchild model FC-2W was basically similiar to the standard FC-2 but was modified somewhat to meet the latest demands imposed by the rapid increase in cargo hauling by air. In it's basic form, it was still a 5 place high wing cabin monoplane, with a strut braced semi-cantilever wing that could also be folded back as on the earlier Fairchilds. In the process of redesign and modification, this model lost the characteristic "pinch-back" fuselage of the early type and also the familiar lower side-windows at the pilot's station. It was now powered with the 9 cylinder Pratt & Whitney "Wasp" engine of 400 to 450 h.p. The wing span and total area was increased somewhat to handle a larger payload and the added power of the "Wasp" engine gave it a substantial increase in performance and general utility, which was found necessary in certain types of "bush flying", especially in Canada. In view of their performance record, it was being generally conceded by now that the "Fairchild" monoplanes were designed and built for rugged service, to carry a profitable payload anytime and anywhere, and they would haul just about anything that would fit into the fuselage! It has been known, that if it was too big to go in, then it was just strapped onto the side of the fuselage, and that was that.

The type certificate number for the model FC-2W was issued in December of 1927 and 3 airplanes of this type were built in that year. Early in 1928, an FC-2W was giving Mexico it's first air-mail service, operated by a Mexican air-line that soon became a part of the rapidly expanding system of the Pan American Airways. In April of 1928, the "Bremen", a German "Junkers" monoplane that was on an east-west crossing attempt of the Atlantic Ocean, had crash-landed on Greenley Island near the wilds of Labrador. An FC-2W, shown here, was flown by "Duke" Schiller, a very noted Canadian "bush pilot", and it was the first to reach the crash scene and offer help. In June of 1928, John Henry Mears and Capt. Chas. B. D. Collyer circled the world in an FC-2W, the "City of New York" shown here, in 23 days and 15 hours. Traveling by airplane, train, automobile, and steamer, they beat the previous record of 28 days set by Linton Wells and Edward Evans in 1926. To continue this chronicle of romantic achievements, the Bell Telephone Labortories had an FC-2W that played a very important part in the development of two-way radio communications. This must surely suggest that the Fairchild FC-2W performed a varied and very active service. Manufactured by the Fairchild Airplane Mfg. Corp. at Farmingdale, Long Island; a division of the

Fig. 78. Fairchild FC-2W, powered with Pratt & Whitney "Wasp" engine. This particular plane circled the world in 23 days and 15 hours.

Fairchild Aviation Corporation.

Listed below are specifications and performance data for the "Wasp" powered Fairchild model FC-2W; wing span 50', chord 84", wing area 313 sq. ft., airfoil Goettingen, length 31' height 9', wts. as landplane, empty 2418, useful load 2182, payload 1000, gross wt. 4600 lb. Wts. as seaplane with "Fairchild" pontoons, empty 2770, useful load 1830, payload 650, gross wt. 4600 lb. Performance as landplane; max. speed 140, cruise 115, land 52, climb 900, ceiling 15,500 ft. Performance as seaplane; max. speed 134, cruise 110, land 52, climb 850, ceiling 15,000 ft., gas cap. 160 gals., range approx. 1000 miles.

The fuselage framework was built up of welded steel tubing, faired to shape and fabric covered. The wing framework was built up of spruce box-type spars and wood truss-type ribs, also fabric covered. The wings could be folded back in a few minutes without any disconnecting of fuel lines, control rods, or cables, this was very advantageous in the north country when a plane had to be hurriedly put under cover due to approaching storm. Inertia type engine starter, metal propeller, and wheel brakes were standard equipment. For the next development in the Fairchild "FC" type, see chapter for A.T.C. #61 in this volume for a discussion of the FC-2W2.

Fig. 79. The first to reach the crash-scene of the "Bremen", was "Duke" Schiller in an FC-2W.

Fig. 80. An FC-2W in Canadian service, note the sizeable stack of mail about to be loaded.

A.T.C. #21
(12-27)
THE "OX-5 SWALLOW"

Fig. 81. 1928 "OX-5 Swallow", Dale "Red" Jackson in foreground. Entrant in 1928 Air Races.

In a discussion amongst a group of old-timers, sooner or later the name of "Swallow" is bound to turn up and though opinions may vary somewhat on one thing and another, as they often will, it is generally conceded by all that the introduction of the "New Swallow" biplane formented a great change in early American aviation. Judging from events as they took place, we might then say that this airplane was the best thing that ever happened to stimulate new interest in civil aviation. The OX-5 powered "New Swallow", which was first introduced early in 1924, should certainly be honored with a certain reverance and must surely be listed as America's pioneer "commercial airplane". It was the first new and practical approach to the light commercial type airplane and the first type that dared to break away from the many old concepts of airplane design that had held sway for too many years. The "New Swallow" design was a bold step forward in this type of airplane and it thoroughly convinced the many skeptics that an OX-5 engine could easily carry three persons and still render a good performance; and it also helped to prove that a biplane would certainly hold together without resorting to the outmoded though still quite customary "two bay" wing cellule, with all it's superfluous struts and wires.

In the early "twenties" there were unmistakable signs of a trend slowly developing to get away from the types like the old beloved "Jenny" and the obedient "Standard", which had been quite faithful to all and had served their purpose well but there was now present a restless yearning to overcome the many short-comings so apparent in these war-bred training airplanes. Most of the early commercial design efforts in this respect turned out to be little more than "cleaned up Jennies", and this would even include the 3 place "Laird-Swallow", a design that goes back to 1919 and was termed the first commercial airplane and later as the ancestor of the "New Swallow" of 1924. So with all due credit to the design, we should say that it was left to the "Swallow" to blaze a new trail across the face of aviation for itself and for the many new and interesting designs that blossomed forth just shortly after.

The original "New Swallow", shown here, was designed and developed by Lloyd Stearman with the enthusiastic and very able assistance of Walter Beech and J. M. "Jake" Moellendick. Upon it's introduction in the spring of 1924, which was the year of the awakening, it caused considerable joyful comment among the flying-folk and it became an almost immediate success; 26 of these airplanes were built and sold in the first 6 months! Sales held up very well during the period of 1924-25 but the resulting success of the

"Swallow" fired many an imagination and a great rash of new designs began coming out, as a consequence this tended to offer a good amount of competition and a certain amount of buyer hesitance. Many were waiting to see what would come out next!

Three "OX-5 Swallows" were entries in the Ford Air Tour for 1925, they were flown by Earl Rowland, who gained fame with "Cessna" a few years later, by John Stauffer, and by "Jake" Moellendick the "maestro" of the Swallow organization. This was the first of the "Ford Air Tours" and no scoring was kept, but in spite of that, the "Swallows" had made a good showing; possibly overshadowed just a bit by the "Travel Air" team that was headed by the inimitable Walter Beech.

The "OX-5 Swallow" for 1927-28, as pictured here, was redesigned somewhat by Waverly Stearman, brother of Lloyd Stearman. Stearman and Beech had already left the Swallow organization some time ago to develop the "Travel Air", and in turn, Lloyd Stearman had left "Travel Air" to develop and produce the popular "Stearman" biplane; these were times of rapid development in aviation and with it came a certain amount of disagreement and upheaval.

In it's basic form, the 1927-28 "Swallow" was still a 3 place open cockpit biplane and was powered with the Curtiss OX-5 engine. It now had "N type" interplane struts for the wings and the center-section, and the wings were still of equal span; with an aileron on each panel that were still cable operated. The landing gear was somewhat improved, and the engine was cowled in neatly but was easily accessible for inspection and maintenance. Radiator installation was typical and it was shuttered for temperature control. The type certificate number for the "OX-5 Swallow" was issued in December of 1927 and about 100 were reported built in that year; plus 3 that were powered with the "Hisso" Model A engine of 150 h.p., and one "Dole Derby" entry, the "Dallas Spirit". The "Dallas Spirit" was a specially built "Swallow" monoplane that was lost at sea while searching for survivors of the "Golden Eagle" and the "Miss Doran", after it's own participation in the race was abandoned!

The "Swallow" remained quite popular about the country and was well liked by many, it flew well, was quite dependable, and it's pleasant lines were very appealing. The standard color scheme seemed to vary from time to time but for most of the later "Swallows" it was a deep shiny black fuselage and bright orange-yellow wings. Fond recollection brings to mind easily the many, many, times we would wash down the oil-soaked "belly" of a certain "Swallow" in exchange for a short "hop". This was always so thoroughly enjoyed too, because the pilot had the most delightful habit of including a few mild "stunts", and then would beat up the field with a grass-cutting "buzz job" or two before coming in to land! The "Swallow" was a good hand at these "buzz jobs" cause she built up speed quickly, like a falling rock! However, the "Swallow" was not always known to be so vivacious, it was sure a ground-loving airplane with a full gross load on a hot still-air day. The "Swallow" flew slightly heavy, like a big ship, and was uncomfortable and slightly unpredictable in a prolonged "spin", But this was not really considered objectional and besides, a good many other airplanes tended to possess this same characteristic.

After a good production run of about 5 years, the 3 place OX-5 powered "Swallow" was tapering off in production and was finally discontinued in late 1929. But "Swallow" wasn't then out of business by any means, they still had other models in the works but concentrated mostly on the "TP" line, which

Fig. 82. The "New Swallow" of 1924, Stearman and Beech in foreground. This airplane sparked a rash of new designs in the next few years.

Fig. 83. "OX-5 Swallow" flown in 1925 Ford Air Tour by Earl Rowland, this was a preview for the 1926 model.

Fig. 84. OX-5 powered "Super Swallow" for 1926.

by the way, was another "Swallow" first! For a discussion of the "Hisso-Swallow" and the "Whirlwind-Swallow", see chapters for A.T.C. #50 and #51 in this volume. For a descriptive discussion of the "Swallow TP", see chapter for A.T.C. #105.

The Swallow Aircraft Co.'s biggest problem was that it's management seemed to be in a continuous state of flux; the "big wheels" in 1924 were Lloyd Stearman, Walter Beech, and J. M. "Jake" Moellendick. When Stearman and Beech left to organize "Travel Air", Moellendick was left holding the bag, till he affected a reorganization. Later, Victor Roos, a familiar aviation figure around Wichita for years and formerly with the Cessna-Roos Aircraft Co., became Gen. Mgr. Then Roos picked up stakes and left for "Lincoln-Page", it was this sort of thing that left "Swallow" with an almost continuous unsettled household and this had it's telling effects. "Swallow" vehemently resented imitation by others, especially by "Lincoln-Page", and strongly hinted of this feeling in many of their ads. They would most always say, "Imitation is a concession to superiorty"! This brings up a point that has been discussed both pro and con by many pilots sitting around a pot-bellied stove during off-times, when discussions such as this were bound to turn up. It was generally conceded by most that there are only so many ways you can go in the design of an airplane, or selection of a configuration. It stands as reasonable then that there's bound to be some imitation or influence bandied about from one type to another, many cases known will illustrate this; usually this is only vaguely noticeable and certainly not objectionable. The "Swallow" in all models were manufactured by the Swallow Airplane Co. at Wichita, Kansas.

Listed below are specifications and performance data for the OX-5 powered "Swallow"; span upper and lower 32'8", chord both 60", wing area 300 sq. ft., airfoil Clark Y, length 23'10", height 8'11", empty wt. 1447, useful load 753, payload 340, gross wt. 220 lb., max. speed 100, cruise 85, land 40, climb 500, ceiling 12,000 ft., gas cap. 40 gal., oil 4 gal., range 450 miles. The 1924 "Swallow" sold for $3500, in 1925 this was reduced to $2750, the 1927 model sold for $2485, 1928 prices ranged slightly higher and so on. The fuselage framework was built up of welded chrome-moly steel tubing, faired to shape with wood fairing strips and fabric covered. The wing panels were built up of solid spruce spars and wood built-up ribs, also fabric covered. The fabric covered tail-group was also built up of welded chrome-moly steel tubing. The interplane struts were of chrome-moly steel tubing of streamlined section, interplane bracing was of streamlined steel wire.

The "Swallow", like other airplanes of this type, performed various and sundry chores. Chores like passenger-hopping on weekends, student instruction, charter trips and the like, and one was even known to be smuggling liquor across the border from Canada!

A.T.C. #22
(1-28)
CENTRAL STATES, "MONOCOUPE"

Fig. 85. 1927 Central States "Monocoupe" with "Air Cat" engine.

"Monocoupe"...here is a name of enchantment that grew to spell fun and pride to those that knew it well and it has since been a name faithfully endeared in heart and memory by many flying-folk, even to this day. When first introduced during the hustle and bustle of early 1927, it came forth as a refreshing departure from the typical commercial airplane type of this period, and had made an early favorable impression on a portion of the flying public. Without the full realization at first, this type of airplane was destined to become very popular, and this early model of these delightful little machines, pictured here, was to become the forerunner to a whole mess of really good and often outstanding little airplanes.

The Central States "Monocoupe" was a diminutive cabin monoplane of somewhat comical proportions, that seated two side by side in chummy comfort, and for want of a better powerplant, was somewhat reluctantly powered with the "phlegmatic" 6 cylinder Anzani engine of some 60 h.p.; some were powered with the new and promising 5 cylinder Detroit "Air Cat" engine. The "Air Cat" was an air-cooled radial engine of some 65-75 h.p., a design by Glenn Angle of Detroit. It was introduced to the trade with high hopes of success, of course, but sadly enough it was short-lived and soon fizzled out. The Detroit "Air Cat" was a development from the former "Rickenbacker" engine, that was introduced in 1926 and also designed by Glenn Angle.

This first series of the "Monocoupe" type was built in Davenport, Iowa by the Central States Aero Co. Inc. and was designed, developed, and carefully nurtured into being by one Don A. Luscombe; with an able assist by Clayton Folkerts. Don Luscombe was a revolutionary, and talented designer, who even later was often suspected of having access to a magic "crystal ball"! But naturally, his apparent knack in airplane development was just careful analysis and good judgement. Don A. Luscombe was at the helm of "Monocoupe" for a good number of years and in more recent years, designed and developed some exceptionally good airplanes under the well known "Luscombe" banner.

Fig. 86. Central States "Monocoupe", ship shown is the original prototype version with Detroit "Air Cat" engine.

Fig. 87. Three Central States "Monocoupes" of 1927, all are powered with the "Air Cat" engine.

After the usual amount of struggle and growth, the Central States Aero Co. Inc. became the Central States Aircraft Co. and sometime later in 1928, the firm was re-organized with new capital and the Mono Aircraft Co. of Moline, Ill. was formed. Still in search of a suitable and compatible powerplant for the "Monocoupe", they tried a version powered with the "Siemens-Halske"; a 5 cylinder engine of some 70 h.p. This little air-cooled radial engine was of German design and manufacture, and it was indeed a commendable powerplant but was not yet the answer that was being sought for the "Monocoupe". A 4 cylinder in-line British "Cirrus" engined version was also hopefully tested, but it was not until the advent of the 5 cylinder "Velie" engine that the "Monocoupe" was considered a right combination, and took off to any degree of national popularity. This "Velie" engined combination was the "Monocoupe 70", see chapter for A.T.C. #70 in this volume.

Central States reported the manufacture of some 22 of the early "Monocoupe" type, some were powered with the Detroit "Air Cat" engine, and some were powered with the 6 cylinder two-row "60 horse Anzani". One of this type (Monocoupe) was reported in Canadian service, fitted with an A-S "Lynx". A Central States "Monocoupe" was flown by Vern L. Roberts in the 1927 Air Derby from New York to Spokane, but was plagued with various troubles and did not finish. Vern Roberts came into considerable fame with the "Monocoupe" type in the next few years. The type certificate number for this early series was issued in January of 1928. As mentioned here before, this early "Monocoupe" type was introduced locally with scant but noisy fanfare in early 1927 and became the first light cabin monoplane to be certificated for manufacture. The roomy cabin featured the companionship of side by side seating and ample performance with good economy set a pattern for many to shoot at; many had tried, but none really had the downright friendly charm of the little "Monocoupe".

Listed below are specifications and performance data for the Central States "Monocoupe" powered with the 60 h.p. Anzani; wing span 30', chord 60", wing area 143 sq. ft., airfoil Clark Y, length 19' 9", height 6'3", empty wt. 700, useful load 475, payload 185, gross wt. 1175 lb., max speed 95, cruise 80, land 38, climb 600, ceiling 8000 ft., gas cap. (2 tanks) 20 gal., oil 2 gal., approx. range 400 miles. Price at the factory for "Anzani" powered model was $2375. The following weights were of the "Air Cat" powered version; empty wt. 749, useful load 475, payload 185, gross wt. 1224 lb. All other specs were comparable and performance was somewhat improved. The "Siemens" and "Cirrus"

Fig. 88. 1927 "Monocoupe" that had found it's way into Canada.

powered versions would also be comparable, or at least within the normal 5% tolerance allowed.

The fuselage framework was built up of welded 10225 steel tubing, lightly faired to shape and fabric covered. The engine mount of chrome-moly steel tubing was detachable by removal of 4 bolts. The wing framework was built up of solid spruce spars that were routed to an "I section", with spruce and basswood built-up ribs; the framework was built up as one piece but the spars were not of continuous length, they were spliced together in the center. A 10 gal. fuel tank was mounted in the wing, one each side of the fuselage. The fabric covered tail-group was built up of welded steel tubing, all controls were cable operated. The landing gear was of the axle and spreader-bar type, rubber shock-cord was used for snubbing the loads, tread was 57". The landing gear and wing struts were of round chrome-moly steel tubing, all exposed tubes were faired with a welded steel tube frame that roughly formed a streamlined section, this framework was fabric covered. Individual seats in the cabin were slightly staggered to offer more shoulder room, the seats and cabin were upholstered and the enormous windows were of pyralin. The cabin entry door was generally on the right side but could be either side, on order. Some liked to fly the throttle with the right hand, so they prefered the door on the left. This feature made them into "left hand" and "right hand" Monocoupes! A "Berryloid" finish of various colors was optional. It might be appropriate to relate that the very first "Monocoupe" was ready for covering in August of 1926 and was personally doped and painted by never-tiring "Tom" Colby; Colby was Mgr. of the Aviation Div. of Berry Bros. Paints and no doubt was in no small way responsible for the popularity of the brilliant "Berryloid Finishes".

A.T.C. #23
(4-28)
BOEING "FLYING BOAT", B-1D

Fig. 89. Boeing B-1D with Wright "Whirlwind" engine, the only example of this type.

It is interesting to note that most all of Boeing's early efforts in airplane design, starting way back in the year of 1916, were either float-seaplanes or flying-boats. No doubt, this was greatly influenced by Boeing's location at Seattle, Wash., an area where there's smooth water aplenty. The company was first formed by W. E. "Bill" Boeing and Conrad Westervelt as the Pacific Aero Prod. Co. in 1916 and introduced it's first airplane, the "B. and W" shown here, in November of that year. It was an open cockpit biplane on twin floats and had a 125 h.p. Hall-Scott engine. Their first development of the "B-1" type was built a few years later and it was a large biplane of the "flying boat" type; it had seating for three in an open cockpit arrangement and was powered with a 6 cylinder Hall-Scott engine of 200 h.p. The Hall-Scott engine might be considered as the grand-daddy of the "Liberty 12", because it was used as a basis for the "Liberty" design. A later development of the Model B-1, shown here, was also a biplane of the flying-boat type and it now seated 4, also in an open cockpit arrangement. It was a somewhat larger "boat" and was powered with a 12 cylinder "Liberty" engine of 400 h.p.

The Boeing model B-1D was built under this certificate number which was issued in April of 1928 and was the first of the flying-boat type to be certificated for manufacture. Despite continuous modifications to this series up to this point, this model was also a biplane of the flying-boat type and it seated four, now in the comfort and protection of an enclosed cabin; the hull design was somewhat modified over the earlier models and it was powered with a 9 cylinder Wright "Whirlwind" J5 engine of 220 h.p. The engine was mounted in a nacelle between the wing panels in a "pusher" fashion, thus getting the "prop" away from the cabin section and minimizing the cause for danger. The B-1D performed well for an airplane of this type and proved to be an excellent boat. It was used for carrying air-mail, cargo, and occasional passengers in the Pacific north-west and in Canada where rivers, lakes, and large bodies of water were convenient and plentiful.

Gone from the scene now, almost entirely

Fig. 90. The 1916 "B & W", Boeing's first airplane.

Fig. 91. Early Boeing B-1 flying boat, powered with "Liberty" engine. This ship flew mail from U.S. to Canada.

in this latter day, is the romance and bountiful pleasures that were experienced with the "boat type" of airplane. A versatile type that tends to combine the sport of boating with the convenience and pleasures of flying. In certain parts of the country that abound in lakes and rivers, it offered a multitude of sporting ventures and also instilled the comfortable feeling to all aboard that a nearby haven always exists in the event of trouble that called for a hasty set-down. Flying-boat, seaplane, and amphibian-airplane pilots will quite well assure you that there is nothing quite like like flying off of water!

The type certificate number for the model B-1D was issued in April of 1928 and it was built by the Boeing Airplane Co. at Seattle, Wash. Two of the B-1D type were reported built, one was built with a "Whirlwind" engine for Percy Barnes, and the one built for Western Canada Airways had a "Wasp" engine, both are shown here.

Listed below are specifications and performance data for both versions of the Boeing model B-1D; span upper and lower 39'8", chord both 79", wing area 466 sq. ft., airfoil "Boeing", length 30'10", height 12', empty wt. 2088, useful load 1155, payload 675, gross wt. 3243 lb., max. speed 110, cruise 95, land 55, climb 750, ceiling 12,000 ft., gas cap. 40 gal., range 300 miles. Weights and performance figures above are for "Whirlwind" powered version. Figures for the P & W "Wasp" powered version were the same except for the following; empty wt. 2588, useful load 1155, payload 675, gross wt. 3743 lb., max. speed 125, cruise 105, land 57, climb 1,000, ceiling 15,000 ft., gas cap. 40 gal., range approx. 175 miles. The hull framework was built up of spruce and was covered with mahagony veneer. The wing panels were built of solid spruce spars and spruce built-up ribs, also fabric covered. Wings were wired for lights. Fuel was carried in the upper wing and the oil tank was in the engine nacelle. All controls were cable operated. The next development of the Boeing B-1 type was the "Wasp" powered B-1E, see chapter for A.T.C. #64 in this volume.

Fig. 92. Boeing flying boat of the B-1D type, with P & W "Wasp" engine, built for service in Canada.

A.T.C. #24
(1-28)
STINSON "DETROITER", SB-1

Fig. 93. A J5 powered "Detroiter" SB-1, flying air-taxi with Wayco Air Service of Detroit.

Edward A. "Eddie" Stinson, master aviator, proudly introduced the first of these "Detroiter" biplanes on a flight from Selfridge Field, Michigan, on Jan. 25 in 1926. It was a trim, fully enclosed cabin biplane with seating for 4 and was powered with a Wright "Whirlwind" of the popular "J4 series". This airplane, shown here, was the culmination of much planning, and with the gracious help of Alfred Verville, it was planned to incorporate some very advanced ideas for these times. Fred Verville, generous of heart and willing to share his tremendous experience and knowledge in aircraft design, helped "Eddie" Stinson through the rough spots in the design of this first "Detroiter". Among some of the more outstanding features built into this airplane, were individual wheel brakes, an emergency parking brake, an electric engine starter to eliminate "propping" by hand. The fully enclosed cabin had a fairly efficient exhaust-manifold type heater which kept it reasonably cozy inside. "Eddie" just loved to show the "Detroiter" off and spent considerable time demonstrating it's features and abilities. Though it was a cold winter in Michigan, he often flew in shirt-sleeves to prove his ship's comfort and utility. Needless to say, "Eddie's enthusiasm and his countless demonstrations were effective and it went over big! Quite a few men of "big business" were favorably impressed and became interested in the airplane's future. Shortly thereafter, the Stinson Airplane Corporation was formed in Detroit to begin it's manufacture. Stinson acquired buildings for the Northville plant (a suburb out of Detroit) in May of 1926, and the first production model was produced in August. With a few more ideas always up his sleeve, Stinson had immediately laid plans for an improved model, so in Aug. of 1926 a new "Detroiter" biplane was introduced; with some modifications that were definite improvements. The new model had a much deeper and better faired fuselage with an improved tail-group; the powerplant was still the reliable J4 and all-round performance was a good bit better.

This model soon proved itself popular as a passenger and mail carrier on some of the early air-lines such as Florida Airways and Northwest Airways, and also served double duty by hauling passengers and all sorts of cargo with Noel Wien in Alaska and Patricia Airways in Canada, to name a few. Northwest Airways started scheduled passenger service with their "Detroiters", one is shown here, in July of 1927 and were of the first to offer air-travel in cabin comfort in the U.S. The "Wayco Air Service" formed by Ed Schlee and his brother, were operating an air-taxi-service out of the Detroit area in May of 1927, using two "Detroiter" biplanes; one of

Fig. 94. "Eddie" Stinson took great pride in his first "Detroiter", which first flew in Jan. of 1926.

these is shown here. The SB-1 was also very popular as a personal transport plane and was used by numerous business executives.

Two "Detroiters", only slightly modified, were used by Geo. Hubert Wilkins on his Arctic exploration expedition of 1927; after many successful sorties, one crash-landed on the Arctic ice and had to be abandoned, the other one was later sold in Alaska for "bush flying" duty. A "Detroiter" cabin biplane was flown by "Eddie" Stinson in the 1926 Ford Air Tour and finished in 3rd place amongst a stellar field of tough competition.

One of the illustrations pictures here a classic incident that shows the first "Detroiter" biplane crashing through a hot-dog stand, this was the incident that motivated the perfection of an emergency parking brake! As the story goes, "Eddie" was forced to "prop" by hand so he left his passenger inside to mind the throttle; when the engine started off with a roar, the excited "throttle-watcher" instead of reducing the r.p.m., shoved the throttle to wide-open and that's when plane met the hot-dog stand! "Eddie" was out a good propeller and some cash for the damages to

Fig. 95. View of incident that prompted the perfection of a parking brake on the "Detroiter".

the stand. Needless to say, a parking brake was rigged up very soon afterwards.

The 1926-27 model of the SB-1 is pictured here in various views, a few of the later type were powered with the new "J5 Whirlwind" of 220 h.p. About 19 of the "Detroiter" biplane were built in 1927; from August 1926 some 22 were built and sold in less than a year. The type certificate number for the "Detroiter" SB-1 was issued in Jan. of 1928 but this was more or less a token gesture because the "Detroiter" biplane had already been discontinued in favor of the "Detroiter" monoplane by this date. Production of the "Detroiter" biplane was discontinued in June of 1927. For accounts of the Stinson "Detroiter" monoplane, see chapter for A.T.C. #16 in this volume. The "Detroiter" biplanes, all except the first one, were manufactured by the Stinson Aircraft Co. at Northville, Mich.

Listed below are specifications and performance data for the "Whirlwind" powered Stinson "Detroiter" biplane model SB-1; span upper and lower 35'10", chord both 63", wing area 350 sq. ft. airfoil U.S.A. 35B, length 28'10", height 10'3", empty wt. 1700, useful load 1200, pay load 600, gross wt. 2900 lb.,

Fig. 96. A Stinson "Detroiter" biplane in Canadian service.

Fig. 97. One of the earliest examples of the SB-1 "Detroiter", in service with Florida Airways.

Fig. 98. Northwest Airways were of the first to offer air travel in cabin comfort, scheduled flights began July 1927.

max. speed 118, cruise 100, land 45, climb 800, ceiling 13,500 ft., gas cap. 70 gal., range 600 miles. The following wts. were given for the later type that was powered with the J5 engine; empty wt. 1815, useful load 1465, payload 800, gross wt. 3280. The performance remained about the same with the possible exception of a landing speed of 48, and a climb, of about 750. The "manufacturers performance figures" differed somewhat from those shown, they were inclined to be optimistic in most instances. Price at the factory averaged around $11,000.

The fuselage framework was built up of welded chrome-moly steel tubing, faired to shape and fabric covered. The wing panels were built up of spruce spars and wood built-up ribs, also fabric covered. The fuel supply of 70 gal. was carried in two tanks that were mounted in the upper wing. The fabric covered tail-group was built up of welded steel tubing, the fin was ground adjustable and the horizontal stabilizer was adjustable in flight. Wheel brakes, metal propeller, and engine starter were standard equipment. The SB-1 was tested with pontoons as an experiment, see illustration.

A.T.C. #25
(1-28)
MAHONEY-RYAN "BROUGHAM", B-1

Fig. 99. Well pleased with "Ryan" performance, "Lindy" picked a "Brougham" for his personal airplane.

"Ryan" history is so colorful and had such an influence on early aviation that it would take much more space than can rightly be afforded in this type of coverage; therefore, somewhat reluctantly, we will have to review it lightly. For early "Ryan" activities we might go back to the years of 1924-25 when most of their "manufacturing" consisted of the redesigning and rebuilding of the war-surplus "Standard" biplane of W.W. I vintage. Modifying these into cabin transports that seated a pilot and 4 passengers in a comfortable fashion. One of these is pictured here. These planes were redesigned and worked out

Fig. 100. The "Ryan-Standard" making daily flights from San Diego to Los Angeles and return in 1925.

Fig. 101. Ryan M-1, first plane for Pacific Air Transport, which is now considered as "Old No. 1" in the annals of United Air Lines history.

by Hawley Bowlus, and a few were used regularly by the Ryan Air Lines on their daily run from San Diego to Los Angeles and return; a very noble service and a daring effort for these early times. The Douglas "Cloudster", which was Donald Douglas' first airplane, was also redesigned and rebuilt in the "Ryan" shops into a cabin transport that seated twelve passengers and a crew of two, this was quite an astonishing feat for an airplane that was powered with only a "Liberty 12" engine of 400 h.p.

The first "all Ryan design", so to speak, was a trim little high wing monoplane that was called the "M-1" and was introduced early in 1926. It was a 3 place open cockpit monoplane that was basically of the "parasol" type, and was powered with a Curtiss OX-5 engine of 90 horse. Later modifications to this model, included such various engine installations as the "Super Rhone", the "Anzani" and the "Hisso", the Quick Radial and the Menasco-Salmson, and also the Wright "Whirlwind" J4 engine of 200 h.p. The J4 powered model M-1, pictured here, was of the type used by the Pacific Air Transport, which was organized by Vern Gorst and trail-blazed by T.Claude Ryan, on their projected Los Angeles to Seattle air line, C.A.M. #8. Contract Air Mail #8 was the longest and the toughest of the early routes, a struggling but successful pioneer effort that was later to be absorbed into the vast "Boeing System".

A standard model M-1 powered with a J4 engine was flown in the 1926 Ford Air Tour by Vance Breese and finished in 8th place. Later on in 1926, the next design to be developed was the model M-2 and it was called the "Bluebird"; experience gained through operation of the M-1 dictated requirements found desirable for this new design, so it was developed as a fully enclosed cabin monoplane that seated a pilot and 4 passengers. The standard powerplant installation was the "Hisso" engine, obviously to keep the price down, but "Whirlwind" power was also available on request. The "Bluebird" design was then modified somewhat and later renamed, thus it became the pattern for the famous "Brougham" type, which first came out as a prototype with a "Hisso" engine.

The first "Whirlwind-Brougham" was being built in the shop when Ryan Aircraft contracted to build the "Ryan N.Y.P.", which was to be the famous "Spirit of St. Louis" that Chas. Lindbergh had flown on his solo-flight across the Atlantic Ocean from New York to Paris, in May of 1927 This

Fig. 102. Lindy's "Spirit of St. Louis" which flew across the Atlantic Ocean and sparked a re-birth in aviation the world over.

immortal flight had electrified the whole world and had literally put aviation back on it's feet, including Mahoney-Ryan Aircraft! So in a way, there were strong ancestral ties between the "Spirit of St. Louis" and the standard production "Brougham". As pictured here in the various views, it was a 5 place high wing cabin monoplane and was powered with the Wright "Whirlwind" J5 engine of 220 h.p.; the earlier models swung a Hamilton wood "prop" but later models used the new "Standard Steel" metal propeller. It first had the "shock-cord" sprung landing gear and the small tail-group like the N.Y.P.; and the burnished swirls on the metal of the cowling, prop spinner, landing gear fairings, and wheel discs, was long a familiar feature of the Ryan "Broughams". This burnished metal design was an ingeneous method devised to hide the waves and the bumps on the hand-hammered cowling of the "N.Y.P." and it looked so good, it was adopted as a standard feature on all production "Broughams"! As we can plainly see, the B-1's resemblance to the "Spirit" was quite apparent, it was trim and lady-like both in appearance and in it's behavior, and was capable of a very good performance. For some time, the "Brougham" B-1 basked in the fame of "Lindy's" plane and quite often was refered to as the sister-ship of the "Spirit of St. Louis". After the N.Y.P. was built, and production was resumed on the "Brougham" type, the first B-1 to come off the "line" went to Frank Hawks, later of speed-dash fame, who flew it in the 1927 Air Tour and finished in 8th place. Later in the year, Frank Hawks flew the B-1 in the National Air Races for the Detroit News Trophy, he came in first for speed and 3rd for efficiency. "Pop" Cleveland flew a B-1 to 8th place in the 1927 Air Derby from New York to Spokane.

Lindbergh's oceanic flight certainly gave "Ryan" a much needed shot in the arm and soon they had production up to three airplanes a week, with a greatly increased staff of people. The type certificate number for the "Brougham" model B-1 was issued in January of 1928 and it sold for $12,200 at this time; the price went down later as the amount of production went up. A few of the model B-1 type were also built as float seaplanes under a Grp. 2 certificate numbered 2-18 that was issued in July of 1928. Three "Broughams" were flown in the 1928 National Air Tour, which was actually the former Ford Air Tour, finishing in 7th and 9th place, and one finishing in 20th place. This straggler was flown by the ever-popular E. W. "Pop" Cleveland of "Aerol Struts" who incidently took delivery on the third "Brougham" to come off the line; the other two ships in the tour were flown by Al Henley and Vance Breese.

Most of the staff at "Mahoney-Ryan" back in these days, read like a "who's who" of aviation; such memorable names as B. F. "Frank" Mahoney, Hawley Bowlus, Don Hall, Fred Rohr, Douglas Corrigan, J. J. "Red" Harrigan, and others. Everyone later becoming a prominent figure in the make-up of aviation history. In about October of 1928, Frank Mahoney, who had been sole owner of "Ryan" since Claude Ryan had sold out his interests just before the N.Y.P. was to be built, had decided to quit and sell out to a St. Louis group. The plant was then relocated on Lambert Field just outside of St. Louis, Mo. At the new plant, production was resumed on the newly improved model B-3, and later with the B-5.

Another famous Ryan record-breaker was the "City of Fort Worth", a Ryan B-1, and a beat-up one at that, which set an endurance record of over 172 hours in the air by refueling in flight. 172 hours is just over a week! The refueling tanker-plane was also a B-1 "Brougham". The "Fort Worth" was flown by Jim Kelly and Reggie Robbins, their record was set in May of 1929 but it had a short life, it was beaten the following month!

Bridging all the years that followed, we come to 1956; two aged and disintegrating "Broughams" had been found and were rebuilt into replicas of the famous "Spirit of St. Louis". These two old hulks were being rebuilt by Paul Mantz in California for a "movie" to be filmed about Chas. A. "Lindy" Lindbergh and his memorable ocean flight to Paris, France almost 30 years previous! Jimmie Stewart, a fine actor, played the part of "Lindy", and quite convincingly too. A third replica was also built, by Jimmie Stewart and his associates to be used as a spare in the filming of various scenes. Seeing these "replicas" in person, was almost like re-living a part of the past, bringing back to mind the day we actually saw the original "Spirit of St. Louis" some 30 years ago!

Just as a matter of record, the first "Brougham" to be built was serial #28, and the

N.Y.P. was #29, the "City of Fort Worth" was serial #52, Lindbergh's personal "Brougham" that he toured the country with in 1928 was #69 and listed as a B-2 type, the Mantz replicas were serial numbers #153 and #159, the 3rd replica was #163, and the last "Model B-1" to be built at San Diego was #178. In all, some 150 Ryan "Broughams" were built at San Diego, Calif., The manufacturer during this period was the B. F. Mahoney Aircraft Co.; earlier "Ryans" were built by the Ryan Air Lines, Inc. also at San Diego and headed by T. Claude Ryan.

Listed below are specifications and performance data for the "Whirlwind J5" powered Mahoney-Ryan "Brougham" model B-1; wing span 42', chord 84", wing area 270 sq.ft., airfoil Clark Y, length 27'9", height 8'9", empty wt. 1870, useful load 1430, payload 800, gross wt. 3300 lb., max. speed 125, cruise 105, land 49, climb 800, ceiling 16,000 ft., gas cap. 83 gal., range 700 miles. Price at factory was $12,200 and later reduced to $9700. The fuselage gramework was built up of welded chrome-moly steel tubing, faired to shape and fabric covered. The wing framework consisted of solid spruce spars that were routed to an "I Beam" section, ribs were built up of spruce and plywood and the wing was fabric covered. Fuel tanks were in the wing root on either side of the fuselage. The fabric covered tail-group was also built up of welded chrome-moly steel tubing, the fin was ground adjustable and the horizontal stabilizer was adjustable in flight. The first few of the early "Broughams" had a shock-cord sprung landing gear like the N.Y.P., but that was soon changed and most of them used the famous "Aerol Struts", with Bendix wheels and brakes. Interior was quite neatly equipped with wicker seats of the bucket and bench type.

Fig. 103. The first production Ryan "Brougham" B-1, flown by Frank Hawks in 1927 Ford Air Tour to 8th place.

A.T.C. #26
(2-28)
"SIEMENS-WACO", MODEL 125

Fig. 104. The "Waco-Siemens" with 9 cylinder Ryan-Siemens engine of 125 b.p.

The likeable "Siemens-Waco", also later known as the Model 125, was more or less a typical "Waco Ten" type and was only faintly disguised by the neat installation of a 7 cylinder or 9 cylinder "Ryan-Siemens" engine that was up front. The "Siemens" engines at this particular time were under license to the Ryan Aeronautical Co. in San Diego for distribution in the U.S.A. from the Siemens-Halske Co. of Germany. These engines were exceptionally smooth running and quite reliable, and were becoming somewhat popular in this country because there was a definite need for a good reliable "air-cooled radial" in this power range. American engine manufacturers were painfully slow in getting something of this sort out on the market. A similiar need in the lower power range also helped foster the somewhat dubious popularity of the various small "Anzanis", and the like. Though perhaps it was fortunate indeed that we had some of these engines available to us at that time to help bridge the gap. As some will recall, the temperament of some of these small "foreign radials" often left much to be desired.

The "Siemens-Waco", as pictured here, first appeared late in 1927 and was typical to the standard model "Ten" in most all respects except for the engine installation which was the 7 cylinder "Ryan-Siemens" (Siemens-Halske) engine of 97 h.p. In it's basic form, this model was also a 3 place open cockpit biplane and it flew and handled well. Soon after, the 9 cylinder "Ryan-Siemens" engine of 125 h.p. was installed for test. This combination also flew and handled well and had surprisingly good performance; as proven by Charlie Meyers when he won the free-for-all stunting contest at the Macon, Ga. air races in March of 1928. This combination of plane and engine, though well worth it, was somewhat costly though and didn't fare too well on the market; usually, the small operator found it almost impossible to warrant such a large investment for his use. Then too, the "Waco-Siemens" would have been a better seller had it not been for the constant nagging promise of suitable American engines about to come out, which held it back. In all, some 21 of this model were built, it's use never reached any great proportions, at least in "Waco" tradition, and it was finally discontinued when the "Siemens-Halske" engines were almost impossible to get due to some labor strife in Germany. By this time, there was some promise shown, and various "American radials" in this power range were beginning to make an appearance.

The type certificate number for the "Siemens-Waco" was issued in February of 1928. Only 4 of this type were built in 1927 and possibly some 17 more were built in 1928.

Fig. 105. "Waco 10" with 7 cylinder Ryan-Siemens engine of 97 h.p., this was probably earliest example of this type.

The "Siemens-Waco" was the first certificated installation of the "Siemens-Halske" engine, although the engine had already been used on numerous experimental prototypes and was also later used on models of the "Travel Air", "Spartan", "Cessna", and "Command-Aire", just to name a few.

As a further experiment in the use of medium-powered air-cooled engines; "Advance" built one test-ship late in 1927 that was powered with the unusual 4 cylinder Fairchild-Caminez engine of 120-135 h.p. But, it never went too far beyond the test stage because of various difficulties encountered with the operation of this engine. A "Caminez-Waco" was flown by M. G. Beard in the 1928 National Air Tour and finished in 21st place. The "Wacos" of this period were still manufactured at Troy, Ohio in ever-expanding quarters, many hundreds had already been built and "Advance Aircraft" was on the threshold of great success. See chapters for A.T.C. #41 and #42 in this volume for the development of the "Whirlwind-Waco" and the Hisso-Waco".

Listed below are specifications and performance data for the "Ryan-Simens" powered "Waco" model 125; span upper 30'7", span lower 29'5", chord both 62.5", wing area 288 sq. ft., airfoil Aeromarine Mod., length 23'9", height 9', empty wt. 1330, useful

Fig. 106. The "Waco-Siemens" was neat and well-mannered.

load 730, payload 340 gross wt. 2060 lb., max. speed 112, cruise 96, land 37, climb 730, ceiling 13,000 ft., gas cap. 37 gal., range 350-400 miles. Price at the factory averaged $5300. For the "Siemens-Waco" that was powered with the 7 cylinder engine of 97 h.p., performance figures and wts. were as follows; empty wt. 1310, useful load 715, payload 320, gross wt. 2025 lb., max speed 105, cruise 90, land 37, climb 580, ceiling 12,000 ft., gas cap. 37 gal., range 400 miles. The construction of the fuselage framework and the wing panels were typical to the "Waco 10". Storage for additional fuel supply was available in a 17½ gallon center-section tank which prolonged the range almost another 200 miles. The 9 cylinder "Ryan-Siemens" engine developed 125 h.p. at 1800 r.p.m. and weighed in at 382 lbs., it's price was $2970 early in 1928.

Fig. 107. The "Ryan-Siemens" made into a neat installation, note unusual valve action and exhaust collector ring.

A.T.C. #27
(2-28)
BOEING, "MODEL 40-B"

Fig. 108. The "Hornet" powered 40-B was the answer to constant demand for faster and better service.

The Model 40-B was the next modification in the Boeing "40 series" and in view of the ever-increasing demands for faster and better service, it was a logical development. Like it s earlier counterpart, the Model 40-A, it was also a rather large biplane carrying a combination payload of mail, cargo, and two passengers. The passengers were seated in an enclosed cabin section and the pilot was seated in an open cockpit; in fact, this arrangement for the 40-B was identical to the earlier model 40-A in all respects because all of the 40-B type were actually a modification of existing model 40-A type. Because the cabin section was isolated from the pilot, some lonesome and ingeneous pilots had devised a speaking tube system to enable them to talk to their passengers and explain and identify the unfolding scenery far down below.

The model 40-B conversion was powered with the larger and more powerful 9 cylinder

Fig. 109. Boeing 40-B, tested with geared "Hornet" engine of 525 h.p.

Pratt & Whitney "Hornet" engine of 500 h.p. The added power increased it's general utility and gave it a somewhat livelier performance. Boeing installed the new "Hornet" engines in 19 of their existing "40-A" type that were used regularly on their own airline-system and thus, as mentioned before, converting them into the Model 40-B type. The modification necessary for this conversion warranted the issuance of a new type certificate number, which was awarded in February of 1928 and then re-issued for some added modifications in July of 1929. Except for a few ships, most of the Model 40 type manufactured up to and during this time were used by Boeing on it's own vast airline system. The trust-worthy "40 type" were of a placid nature and obedient, were well adapted to their assigned work and they were proving themselves to their credit; it's not surprising then that they went into quite a few more modifications before being finally discontinued in favor of more specialized equipment. See chapter for A.T.C. #54 in this volume for discussion of the Model 40-C. The "40 series" were manufactured by the Boeing Airplane Co. at Seattle, Washington.

Listed below are specifications and performance data for the "Hornet" powered Boeing "Model 40-B"; span upper and lower 44'2", chord both 79", wing area 547 sq. ft., airfoil "Boeing", length 33'4", height 12', wheel tread 88', empty wt. 3543, useful load 2536, payload 1436, gross wt. 6079 lb., max. speed 132, cruise 110, land 57, climb 800, ceiling 15,000 ft., gas cap. 140 gal., oil 12 gal., approx. range 550 miles. Price at factory was $24,500. The fuselage framework was built up of welded steel tubing braced with steel tie-rods, faired to shape and fabric covered. The wing panels were built up of spruce spars and wood built-up ribs, also fabric covered. The fabric covered tail-group was also built up of welded steel tubing. All of the engine cowling, cabin doors, and compartment hatch covers were formed of sheet dural. The landing gear was of the typical Boeing cross-axle type, using Boeing "oleo" shock-absorbers and individual wheel brakes. Among the added features were heat for the passenger's cabin and safety-glass windows, an electric inertia-type engine starter and complete night flying equipment, including parachute flares and landing lights were provided. Later, all ships were bonded and shielded for radio installation. At re-issuance of certificate on July of 1929, the following wts. were allowed; empty 3714, useful 2365, payload 1228, and gross wt. of 6079 lb., no change in performance figures.

Fig. 110. Boeing 40-A returned to factory for conversion to 40-B type, after over 1400 hours of service.

A.T.C. #28
(3-28)
LINCOLN-PAGE, LP-3

Fig. 111. The OX-5 powered Lincoln-Page LP-3 of 1928.

"Lincoln-Page" history has been very elusive, so it has been difficult to dig up very much accurate information concerning this "type", beyond what is written here. We do know that the Lincoln-Page LP-3 was a typical 3 place open cockpit biplane and was also powered with the popular Curtiss OX-5 engine of 90 h.p. Comparable to the many other airplanes of this type, it was primarily designed for use by the small independent operators and private owners. It was an airplane of quite familiar lines, it may be unkind to say so, but imitation is very evident and you would certainly have to do a double-take to make sure that you weren't mistaking it for an "OX-5 Swallow", for which it had such a striking resemblance. There might have been some valid explanation for this, but it is doubtful. Lincoln-Page should at least be commended for it's faithful reproduction of the "Swallow's" classic lines.

First introduced in the latter part of 1927, the first production model was flown from the factory in Lincoln, Neb. to Chicago, Ill. by the veteran Ed Heath in Jan. of 1928. The type certificate number for the LP-3 was issued in March of 1928 and it was manufactured by the Lincoln Aircraft Corp. at Lincoln, Nebraska. Lincoln Aircraft previously built the "Lincoln-Standards" that were quite popular around the country in the years of 1923-24-25 and they also built the pint-sized "Lincoln Sport", an interesting little one place biplane that was designed by Swanson. Swen S. Swanson was a brilliant man of great capabilities and extreme modesty. Auggie Pedlar, who was later lost at sea in the "Dole Derby" race to Hawaii, once flew a "Lincoln Sport" off of Main St. in Sioux City, Iowa to demonstrate it's incredible performance and it's versatility. S. S. Swanson later designed such interesting airplane types as the lovable "Arrow Sport", the keen little "Kari-Keen", the stately "Swanson Coupe", the "Fahlin Coupe", and the "Plyma-Coupe".

The Lincoln-Page line of airplanes surely had an interesting ancestral background that dates back to the turn of the "twenties" when the Lincoln Aircraft Co., like many another, was rebuilding and remodeling the war-surplus "Standard" biplane into a more useful version that was called the "Lincoln-Standard". These craft were quite popular in the early "twenties" and were available in a 2 or 3 place open cockpit "sport" type, and a 5 place "cabin cruiser", the LS-5. Starting out originally as Nebraska Aircraft sometime in 1922, it was continuously under the able guidance of Ray Page. Activity at "Lincoln" was always high even during these early times and as was often said, some manner of aviation history was always in the making at Lincoln, Nebraska. Chas. A. Lindbergh, famous for his solo-flight across the Atlantic Ocean to Paris, was a student-pilot at "Lincoln Aircraft" and was being taught the fine art of flying by the venerable Otto Timm, a pioneer-airman of some renown.

Ray Page, another pioneer airman, was the president and founder of Lincoln Aircraft and later had the assistance of Victor Roos who had come over from Swallow Aircraft in Wichita, Kansas, Ray Page as head-man of Lincoln Aircraft in 1925, distinctly saw the handwriting on the wall by now and knew that the new production airplanes such as the "Swallow", "Travel Air", "Waco", and others, would make rebuilt war-surplus airplanes very unattractive as a commercial investment. So with concerted effort he started to get out from under his huge stock of "Standards" by selling cheap and as fast as possible, but even at that he was obliged to build up and sell 40 of the modified "Lincoln-Standards" as late as 1927 and these were about the last of the lot. It was then that they developed the LP-3.

The Lincoln-Page LP-3 was popular to some extent in the western half of the U.S.A. but it was little heard of in many other parts of the country, and no records seem to be available of the number that might have been built. Later, there was a modified version known as the LP-3A and it was powered with the "Hisso" Model A engine of 150 h.p.

Listed below are specifications and performance data for the OX-5 powered Lincoln-Page model LP-3; span upper and lower 32'8", chord both 58", wing area 298 sq ft., length 23'2", height 8'10", empty wt. 1350, useful load 850, payload 370, gross wt. 2200 lb., max. speed 100, cruise 85, land 40, climb 500, ceiling 12,000 ft., gas cap. 40 gal., oil 4 gal., range 400 miles. Price at factory was $2250 less engine, customer could provide engine or buy one from Lincoln-Page stock. The fuselage framework was built up of welded chrome-moly steel tubing, faired to shape with wood fairing strips and fabric covered. The wing panels were built up of solid spruce spars and spruce and plywood built-up ribs, also fabric covered. The fabric covered tail-group was also built up of welded steel tubing, the horizontal stabilizer was adjustable in flight. Tail-skid was first a steel tube with hardened shoe, but was later changed to the spring-leaf type.

Just in manner of discussion, it must be taken into consideration that the listed "cruising range" of an airplane is usually only to be taken as an average figure, because pilot habits at the throttle varied considerably and consequently some got more miles per gallon than others; the difference could often amount to as much as 50 miles on a tank of gas!

Fig. 112. The LP-3 of late 1928 had numerous modifications, note enlarged fin and redesigned center-section cabane.

A.T.C. #29
(4-28)
N.A.S. "OX-5 AIR KING"

Fig. 113. "Air King" Model 28, one of the last examples of the type.

First upon the scene in 1926, the "Air King" biplane had been modified and improved in each successive model, and the 1928 version as pictured here, was the culmination of all the previous efforts and turned out to be a pretty fair airplane. As were all the others, this latest and last version of the "OX-5 Air King" was a 3 place open cockpit biplane and was powered with an 8 cylinder Curtiss OX-5 engine. This last model was also of the "long wing" type, a characteristic feature of all previous "Air Kings", but otherwise it was redesigned entirely and rather extensively. The fuselage framework was now built up of steel tubing, but the joints were not welded, they used special steel clamps to join the tubes. All of the earlier "Air Kings" also used these special clamps to join the tubes into a framework, but were built up of seamless "dural" tubing. This was a method of construction quite similiar to that used by "Matty" Laird for many years.

The 1928 model was faired out to a much better shape with improved cockpits and a redesigned tail-group, the engine was now neatly cowled in, but the engine radiator was still hung up in the centersection; a placement that was typical of every "Air King" model built. The landing gear was also greatly improved and was now of the long-leg "oleo spring" type of fairly wide tread, offering much better ground stability. The wing panels were still typical with the lower panel having the greatest span, workmanship of a better class was now much more in evidence, and all of the center-section struts and outer interplane struts were now of streamlined steel tubing. Earlier models had used "dural" tubing for the struts and these were encased in balsa-wood fairings. With all of the aforementioned improvements, the "Air King" was now a sleek and fairly good looking airplane and somehow seemed vaguely to suggest a blend of the "Swallow", "Waco", and "Eaglerock". There hasn't been much "lore" available, either in print or by word of mouth about the "OX-5 Air King", but it is apparent that performance-wise it was about average for a ship of this type.

Production of the "Air King" type in the first year of 1926 was almost negligible but things had picked up a good deal in 1927, and about 55 were reported built in that year. The greatest portion of these were in use by a few flying-school systems that seemed to have some sort of an affiliation with the manufacturer of these airplanes. The latest and last model of the OX-5 powered "Air King" received it's type certificate number in April of 1928 and was more or less in continuous production until some time in 1929.

In August of 1927, N.A.S. had an "Air

Fig. 114. "OX-5 Air King", showing abundance of wing area.

King" entry in the classic "Dole Derby" that was a race across the Pacific Ocean to Hawaii. It was one of their standard biplane models that was fitted with a "Whirlwind J5" engine and extra fuel tanks; luckily, it was disqualified and not allowed to enter because of insufficient cruising range. It would have run out of gas about 300 miles short of it's destination! Chas. W. Parkhurst of Lomax, Ill. was the eager and willing but very foolhardy pilot.

Later in 1928, the N.A.S. (National Airways System) brought out an interesting 4 place cabin monoplane called the "Monofour". It was a fully enclosed high wing monoplane with deluxe appointments, it was powered with a J5 "Whirlwind" engine and was a real smooth looking and well built airplane. The "Monofour" would have probably sold very well, but things hadn't been going too well at N.A.S., and they were just about ready to give up and quit. A cute twist in some of their ads was almost comical and at least good for a chuckle. To quote, "the Air King is the only airplane that is built according to strict N.A.S. standards". The N.A.S. by the way, was the National Airways System, the manufacturer of the "Air King"! Did they actually believe that someone would be impressed or influenced to any degree with something that was so obvious as that.

Listed below are specifications and performance data for the OX-5 powered "Air King"; span upper 31'2", span lower 34'1", chord both 60", wing area 310 sq. ft., airfoil Clark Y, length 25'5", height 9', empty wt. 1380, useful load 755, payload 340, gross wt. 2135 lb., max. speed 98, cruise 84, land 35, climb 520, ceiling 12,000 ft., gas cap. 35 gal., oil 4 gal., range 380 miles. Price at factory in Lomax, Ill. was $2395. The construction of the fuselage framework was discussed previously. The wing panels were built up of solid spruce spars and wood built-up ribs, fabric covered. The fabric covered tail-group was built up of welded steel tubing. The N.A.S. had later moved from Lomax to Peoria, Ill. and had modified the "Air King" somewhat as a last-ditch stand, among the changes was a fuselage framework of welded steel tubing and a redesigned tail-group.

Fig. 115. Late 1927 "OX-5 Air King", shown here at Ford Airport during 1927 Ford Air Tour.

A.T.C. #30
(3-28)
OX-5 TRAVEL AIR, MODEL 2000

Fig. 116. "Old Elephant Ears", shown here in all it's glory.

"Old Elephant Ears".....here is a descriptive phrase that has become a classic thru' the years, and had become to mean only one thing; a phrase that is sure to cause many an old-timer to smile gently at it's utterance and get starry-eyed. Perhaps he will recall some joyful memories of hours spent with the "Travel Air" 2000, which was just about the best darned "OX powered airplane" that was ever built. Most would be hard put for an explanation of just why; we'd say that the "OX-5 Travel Air" was an airplane that really didn't have any one special or outstanding quality that could be singled out as such, but perhaps it was sort of friendliness that one felt in it's presence and it's eager willingness to do well. Surely, it was known to be ever dependable, it really had beautiful lines, a rugged constitution and a pleasant character too; it had a surprisingly lively performance when called upon to give extra effort but more than that, it was altogether just a down-right lovable airplane! Some will say it is true, and this has also been said about many another airplane and they would probably be right, but of the "Travel Air" it seemed to imply without reservation.

The "OX-5 Travel Air's" versatility marked it as an efficient and obedient work-aday airplane throughout the week and when "spruced up" for the occasion, was still capable of turning into a colorful and awe-inspiring week end "show-off"; all primed to do her bit for the promotion of aviation by "stunting" to draw the crowds, and other chores which usually consisted of hopping passengers to make new friends for the art of flying. The pilot of these times usually turned "aviator" for this occasion and if one can remember back then, was resplendent in his carefully creased olive-green shirt and natty bow-tie, with fawn colored breeches and riding boots; a chamois-skin helmet with a pair of "Seesall" goggles pushed back to just the right angle and smoking a cigarette (borrowed) while standing proudly but ever so nonchalantly alongside his "Travel Air". Truly, this was a sight to behold!

As was the manner and custom during this period, the Travel Air "2000" was also a 3 place open cockpit biplane and was powered with the Curtiss OX-5 engine of 90 h.p., that is, if you could get it to "turn 1450"; the OX-5 was often reluctant to do this! Among the many and varied attributes of "Old Elephant Ears", was it's pleasing configuration that bore such a marked resemblance to the famous "Fokker D7" of W.W. I vintage. Often, but not with malice, they were reffered to as "Wichita Fokkers" and were used in many "movies" portraying the first World War period. The large "balance horns" on the ailerons are what earned them the nick-name that became so intimately known as "elephant ears". True, this wasn't a "Travel Air" exclusive by any means, because there were scores of airplanes that had used this type of aerodynamic balance for the ailerons; on the "Travel Air" it seemed so much more prominent and it fit so well with the rest of the airplane. Another familiar feature on the 2000 were the long "tail-pipes" on each side to extend the exhaust fumes well below the level of the cockpits. For some young-bloods, these pipes were much too quiet and the regular OX-5 exhaust manifold had a rather bad sound; it seems they wanted something more "racy" and more "jazzy" so they would replace all this with what was known as "short stacks". These had a beautiful, sharp staccato sound and seemed to give the "old OX" a little extra power and all was swell unless, you should be tempted to put her in a nice long side-slip, that was usually heading for trouble; because the valves would catch the cold blast and curl up like the brim on a derby hat!

Travel Air Inc. was formed in Jan. of 1925 by Walter Innes Jr., Lloyd Stearman, Walter Beech, and Clyde V. Cessna. The first "Travel Air" to be built was the "Model 1000" as shown here. It was a 3 place open cockpit biplane and was powered with a Curtiss OX-5 engine; it was nearly identical to the later models except for many minor and nearly unnoticeable details. It's most prominent difference was the axle-type landing gear. The model "1000" was introduced and formally announced in March of 1925, being built to the inspired designs of Lloyd Stearman who was ably assisted in it's development by such stalwart men of aviation as Walter Beech, Clyde Cessna, and Mac Short. Every one of them a near-genius in his own right, and who have since become famous names indelible in the recorded annals of aviation. This first "Travel Air" was built with high hopes and loving care in the back part of an old abandoned creamery building, although some accounts say it was an old planing-mill, and some even say that it was just a rented garage. Anyhow, it is needless to say or speculate which was right because that's just about how most all of the aircraft companies got started in these early days! Yet, from this modest start, by the year of 1929, Walter Beech who was by then the sole guiding genius, had one of the largest commercial aircraft plant layouts in the country.

Of the first few "Travel Airs" built, three were participants in the 1925 Ford Air Tour, which was the first of these grueling reliability tours and they did very well; all coming up with near-perfect scores. They were flown in the tour by Walter Beech, "Chick" Bowhan, and E. K. Campbell. The comely "Travel Air" had caught on fast and was doing fairly well for itself in the period embracing 1925 and 1926. They had sold something like 19 airplanes in 1925, some of which were custom-built to customer's specifications; in 1926 they had built something like 46 airplanes and then in 1927-28 the orders for "Travel Airs" poured in like an avalanche! The type certificate number for the OX-5 powered "Travel Air" was issued in March of 1928 and as was reported, a total of about 530 airplanes of various models and types were built in that year. The biggest majority of these were of course the model 2000; incidently, the popularity of the OX-5 powered "Travel Air" held up admirably and never waned, they were continuously built into early 1930.

In view of the fact that so many hundreds had been built, it was not surprising to learn that a good number of these old "elephant ear" type of "Travel Air" are still flying, though most have been modified for special purposes and re-fitted with modern radial or in-line engines; this number includes many "crop dusters" and one old "2000" that had been flying around the country masquerading as a bright red "Fokker D7". Even the original "Model 1000", old #1 itself, is still in existence and has been carefully rebuilt to fly again as sprightly as it did that day some 35 years ago! Interest in reviving the "Travel Air" seems to run very high and we've heard of, and from, many an old-timer who literally swore in good faith that they'd just about give a right arm, or do something equally as drastic to own and fly an "OX-5 Travel Air" again; their devotion for the type went that deep!

When first introduced, the "OX-5 Travel Air" cost $3500 thruout most of the year in 1925; then the price was lowered to $3100 and so on, finally getting down to the low of $2195 in early 1930. The "Travel Air" was manufactured by the Travel Air Mfg. Co. at Wichita, Kansas. Later in 1928, the firm's

Fig. 117. The "Model 1000", trying out it's wings over the Kansas country-side in 1925.

structure was re-organized into the Travel Air Co., in early 1930 Travel Air, like so many others, was feeling the effects of the "crash" and the ensuing "depression"; unhappily, it consented to be absorbed into the relative safety of the mammoth Curtiss-Wright organization for self preservation.

Listed below are specifications and performance data for the OX-5 powered "Travel Air" model 2000; span upper 34'8", span lower 28'8", chord upper 66.75", chord lower 55.75", wing area 297 sq. ft., airfoil "T.A. #1", length 24'2", height 8'11", empty wt. 1335, useful load 845, payload 380, gross wt. 2180 lb., a "2000" had once taken off with a gross load of 2600 lb., which is a useful load nearly equal to it's own weight, this was quite a feat for an "OX-5 powered" airplane. Max. speed was 100+, cruise 85, land 40, climb 550, ceiling 10,000 ft., with just a pilot aboard, some 2000 type have been known to climb to 14,000 ft. and upwards. Gas cap. 42 gal., oil 4 gal., range 425 miles.

The fuselage framework was built up of welded chrome-moly steel tubing, heavily faired to shape with wood fairing strips and fabric covered. The wing panels were built up of laminated spruce spars and spruce and plywood built-up ribs, finished panels were fabric covered. All interplane struts were of streamlined chrome-moly steel tubing, interplane bracing was of streamlined steel wire. The fabric covered tail-group was built up of welded steel tubing, the fin was ground adjustable and the horizontal stabilizer was adjustable in flight. The standard color scheme for the "2000" was a "Travel Air Blue" fuselage and tail-group, with silver wings, although some of the earliest models used various color combinations; black and gold was quite popular. A few of the 2000 type were built as float-seaplanes under a Grp. 2 certificate that was issued in Feb. of 1929. For a discussion on the hairy-chested "Hisso-Travel Air", see chapter for A.T.C. #31 in this volume.

Fig. 118. 1929 Travel Air "2000", this was No. 810, one of the last examples of this type.

A.T.C. #31
(3-28)
TRAVEL AIR, MODEL 3000

Fig. 119. The "Hisso" powered "Travel Air" was a "hairy-chested" performer.

The handsome and dashing, model 3000, was generally referred to as the "Hisso-Travel Air" and though it was a popular airplane too, it's following was a mere handful when compared to the number of 2000 owners. In it's basic form, it was also a 3 place open cockpit biplane and was typical to the model 2000 in most all respects except for the engine installation and a few modifications that were necessary for this combination. The engine in this case was either the "Hisso" (Hispano-Suiza) Model A of 150 h.p. or the hi-compression Model E of 180 h.p., the "A" installation was more often used. The 8 cylinder water-cooled vee-type Hispano engines were also of W. W. I vintage like the beloved "OX-5", and were also available from war-surplus stocks about the country, but they cost a good deal more to buy and cost very much more to operate and maintain. It is most likely for this reason alone that the popularity of this engine was somewhat limited. Still, they did help to fill a need for power in this higher range until the "American radials" got a chance for a good foot-hold and finally took over.

As can be seen in the view pictured here, the "Hisso-Travel Air" was a good looker, it was a lively, hairy-chested, and good performing airplane too and it's resemblance to the famous "Fokker D7" was also used to good advantage in many "movies" of the W. W. I period. Here it always flew somewhere in the background to help fill out the "Baron's Staffel". Some were also performing in "air shows", at a time when "barnstorming" though actually on the wane, was still showing a fair profit, and doing the county-fair circuit each fall was still a great American tradition. The "3000" was known as a pretty fair stunt-ship and certainly a great crowd pleaser, especially when it blasted by at low level with it's "Hisso" roaring at full bore!

Four "Hisso" powered 3000 type were built in 1927 and the type certificate number for this model was issued in March of 1928. Being built thru 1928 and into 1929, some 33 of these were built in all. As pictured here, we can see that the 3000 version was a typical "Travel Air" biplane even down to the familiar "elephant ear" ailerons and some also used the characteristic exhaust "tail-pipes". The very first "Hisso" powered "Travel Air" was actually brought out in 1926, but it was a special custom built design to be used as an air-transport. It was a combination cabin and open cockpit biplane that seated five, it had a cabin section that seated four passengers and the pilot was seated aft in an open cockpit. Later a model of this type was powered with a "J4 Whirlwind" engine and was used in Canadian bush-country

Fig. 120. "Hisso" powered Travel Air, an entry in 1926 Ford Air Tour; one of the earliest examples of this type.

service.

Louise Thaden, a famous and very capable woman-flyer of this period, coaxed her "Hisso" powered "Travel Air" to an altitude of over 22,500 ft. for a new woman's record. That was in December of 1928, and in March of 1929 she set a solo endurance record by staying up for just over 22 hours in that very same "Hisso-Travel Air". It is most likely that Louise Thaden did more to bring the model "3000" before the public for proper recognition of it's merits, than any other one person. Another "believer" in the "Hisso-Travel Air" was D. C. Warren of California who entered his in just about every air race in that part of the country.

Listed below are specifications and performance data for the "Hisso" powered "Travel Air" model 3000; span upper 34'8", span lower 28'8", chord upper 66.75", chord lower 55.75", wing area 296 sq. ft., airfoil "T.A. #1", length 24'3", height 9', empty wt. 1664, useful load 926, payload 500, gross wt. 2590 lb., max. speed 112, cruise 100, land 46, climb 800, ceiling 15,000 ft., gas cap. 42 gal., range 425 miles. The above listed figures are for the "Hisso A" version, figures for the "Hisso E" version were the same except as follows; max. speed 119, cruise 105, land 46, climb 900, ceiling 17,000 ft., cruising range on 42 gal. was only slightly less. The general construction details for all parts of this airplane were typical to all "Travel Air" type of this period, consult previous chapter. For discussion of the "Whirlwind-Travel Air", the Model 4000, see chapter for A.T.C. #32 in this volume.

Fig. 121. A later version of the Hisso-Travel Air: one of the last examples of this type.

A.T.C. #32
(3-28)
TRAVEL AIR, MODEL 4000

Fig. 122. J5 powered "Travel Air" 4000, flown by "Billy" Parker for Phillips Petroleum Co. in 1928.

The "J5 Travel Air" or model 4000, as pictured here in various views, was a glamorous and charming old sweetheart that was easily the center of attraction on most any field, whether at rest in her charming dignity, or performing aloft. All those who remember, will say that she certainly was of a "classy breed" and showed this in her behavior. In it's basic form, the "Whirlwind-Travel Air" was a 3 place open cockpit biplane and was quite typical of the other "Travel Air" biplanes currently built at this time, except for the engine installation and some modifications that were necessary for this combination; it had the added power of a 9 cylinder Wright "Whirlwind" engine of the "J5 series" that was rated at 220 h.p. The extra "horses" gave the "J5 Travel Air" a thrilling and very responsive performance, a ship just made to order for the true sportsman-pilot, especially if he cared to indulge in some "stunt-flying", which usually was a form of aerobatics born of enthusiasm but not particularly concerned with precision. The "4000" was precise and very nimble, and was indeed a great "show-off". Besides being the choice of many sportsman-pilots, to many this was a "dream ship"; to the boys who flew their OX-5 powered "Travel Airs" around patiently and were sincerely wishing that someday they would be "flush" enough or fortunate enough to get the feel of a J5 up front! One enterprising young fellow felt the urge quite strongly and decided to do something about it; he yanked out "the old OX" and put in a "Whirlwind J4" that once had flown in a "Detroiter" biplane. After a certain amount of figuring and very much work, he had his "dream ship" realized!

Some 14 of the "Model 4000" were built in 1927 and it's type certificate number was issued in March of 1928. The "J5 powered" model 4000 went into a number of modifications before the "Whirlwind J6" series came along in late 1929. One of these later versions of the J5 powered 4000 was the specially built D-4000. This ship was built for Paul Braniff and was one of the first examples of the so-called "speed wing" type. For still another version of the "J5 Travel Air", see chapters for A.T.C. #146 and #147. Later on in 1929, there were also many other variations on the "Model 4000" by the installation of various air-cooled radial engines that were available at the time, just about everything from the 5 cylinder "100 horse Kinner" on up thru to the "Whirlwind" J6 of 300 h.p. That is, excepting the models 8000 and the 9000, for a discussion on these see chapters for A.T.C. #37 and #38 in this volume.

Travel Air introduced the "4000 type" early in 1926 and it was also powered with a

Fig. 123. J-4 Travel Air flown by Walter Beech in winning the 1926 Ford Air Tour.

9 cylinder Wright "Whirlwind" engine but of the early "J4 series" that was rated at 200 h.p. This "J4 Travel Air", shown here, was soon put into prominence by Walter Beech, always a real go-getting hustler, who flew it in the 1926 Ford Air Tour and placed it in a solid first place win, with Louie Meister in a B/V "J4 Airster" at his heels in a good second place. The early "Whirlwind-Travel Air" was basically typical to the later ones as built in 1927 and 1928; except for the engine installation, a slight difference in the landing gear and just a few other minor details. The 1928 model of the 4000, as shown here, was "elephant ears" in all it's glory. These airplanes were very well liked and owned by many a prominent figure such as "Billy" Parker of Phillips Petroleum, by Ken Maynard and Wallace Berry, both famous of the cinema, and a host of others. Flying schools, such as the famous "Parks Air College", used it for advanced phases of instruction to train "transport" pilots. A versatile airplane, the "Whirlwind-Travel Air" was a type that will be long remembered.

Listed below are specifications and performance data for the J5 powered "Travel Air" model 4000; span upper 34'8", span lower 28'8", chord upper 66.75", chord lower 55.75", wing area 296 sq. ft., airfoil "T.A. #1", length 23'6", height 9'1", empty wt. 1650, useful load 762, payload 340, gross wt. 2412 lb., max. speed 130, cruise 110, land 45, climb 1200, ceiling 20,000 ft., gas cap. 42-60 gal., range 450-575 miles. Price at factory in 1928 was $9100. The following wts. are for the early "light job"; empty wt. 1400, useful load 1000, payload 600, gross wt. 2400, various performance figures were affected accordingly. Subsequent variations of the "J5 Travel Air" were somewhat heavier, carried substantial increases in payload and therefore operated with heavier gross loads, performance in some instances suffered accordingly. All construction details of this "Travel Air" were typical, except for modifications made necessary to absorb the added stresses of increased power. See previous chapters. An inertia-type engine starter, metal propeller, and wheel brakes were available, the wings were wired for lights.

Fig. 124. Six J5 powered "Travel Airs" lined up at Parks Air College, "ring cowlings" and low pressure tires were later additions.

A.T.C. #33
(4-28)
BUHL "AIRSEDAN", CA-5A

Fig. 125. The handsome Buhl "Airsedan", note spacious cabin and excellent visibility.

The model CA-5A, as pictured here in various views, was the second in the Buhl line of "Airsedans" and was but a deluxe refinement over the earlier CA-5 that was built in 1927. This new ship was a modification in many small details and like the earlier "Airesdan", it's outstanding feature was the small span and area of the lower wings. The "Airsedan" was not yet but was approaching the "sesqui-wing" type of airplane. This was an arrangement reputed, especially by Buhl, to have the distinct advantages of both the monoplane and the biplane; in fact, the "sesqui-plane" was generally reffered to as a "monoplane and a half". This type of wing arrangement allowed a shorter over-all wing span in spite of the generous amount of wing area employed, and allowed for a simple yet robust method of bracing for the wing truss.

In it's basic form, this new "Airsedan", the model CA-5A, was a fully enclosed cabin biplane with ample seating in spacious comfort for five. The powerplant for this model was also the 9 cylinder Wright J5 engine of 220 h.p., and it's performance was a good average for a ship of this size and type. At least one of the CA-5A type broke into the lime-light of nation-wide prominence. This was the "Angeleno" that was flown by Loren Mendell and "Pete" Reinhart in setting an endurance record of over 246 hours by refueling in mid-air. They cruised lazily over So. California while their fuel and other personal needs were tended to by an old Curtiss "Carrier Pidgeon". Their record was set in July of 1929, and it bettered a record set just a week or so previous, but their record didn't last long either, just about 2 weeks, because it was bettered by the "St. Louis Robin" which stayed up over 420 hours! And so went the path of aviation progress. Incidently, the first air to air (plane to plane) refueling flight was accomplished by two "DH-4B" type of the Army Air Service back in 1923, they set a record of 37 hours!

The type certificate number for the model CA-5A was issued in April of 1928, but not very many units of this model were built. The "Airsedan" type was certainly a very good airplane, but it's recognition was rather slow and it hadn't come into any sort of nation-wide popularity as yet. A good sales force could have done wonders with this airplane. The "Airsedan" was manufactured by the Buhl Aircraft Co. at Marysville, Mich., Lawrence D. Buhl was the Pres. of the firm and was a member of the prominent Buhl family of Detroit, Mich. that was very active in various business enterprises. Ettienne Dormoy presided over the engineering department which consisted of four men at most.

Listed below are specifications and per-

Fig. 126. Buhl "Airsedan" CA-5A, taking off from factory field in Marysville, Mich.

formance data for the "Whirlwind J5" powered Buhl "Airsedan" model CA-5A; span upper 42', span lower 32'4", chord upper 72", chord lower 48", wing area 334 sq. ft., airfoil Clark Y, length 27'8", height 8'10", empty wt. 2100, useful load 1600, payload 1000, gross wt. 3700 lb., max. speed 122, cruise 102, land 48, climb 700, ceiling 13,500 ft., gas cap. 70 gal., oil 5 gal., range 650 miles. Price at factory field was $12,500 and then later upped to $13,500. The fuselage framework was built up of welded chrome-moly steel tubing, faired to shape and fabric covered except in the portion surrounding the cabin area which was covered with sheet aluminum panels. The wing panels were built up of the normal spruce spar and spruce and plywood built-rib arrangement which were also fabric covered. Fuel supply was carried in two tanks, one in each wing root flanking the fuselage. The lower wing roots were heavy in cross-section and were built integral with the fuselage to absorb the extra stresses imposed on it by the landing gear, which was fastened directly to the lower wing stubs. The landing gear was modified a bit and was approaching the type similiar to that used on the later CA-8 models. The fabric covered tail-group was built up of welded chrome-moly steel tubing, the fin was ground adjustable and the horizontal stabilizer was adjustable in flight to trim for variations in loading. Inertia-type engine starter, metal propeller, and wheel brakes were standard equipment, wings were wired for lights. For the next development in the "Airsedan" series, the CA-3C "Sport Airsedan", see chapter for A.T.C. #46 in this volume.

Fig. 127. "Airsedan" utility paid off on feeder-routes and charter flights.

A.T.C. #34
(4-28)
LOENING, "WASP-AMPHIBIAN"

Fig. 128. The Loening "Cabin Amphibian" with "Wasp" engine.

At rest or aloft, it must be conceded that the Loening "Amphibians" were truly one of the most striking and easily recognizable configurations of all time. A design that was unusually distinctive because of it's unorthodox layout, possessed of features that surely made them appear rather odd but yet so unmistakably functional. As pictured here in the various views, it is plain to see just why they earned the nick-name of "Flying Shoehorns"!

This particular model, the "Wasp-Amphibian", discussed here and pictured in the title photo, was the first of the Loenings designed primarily as a passenger-carrying transport and the first of the "Amphibians" to be certificated; although many and numerous other models had been built prior to this one. This transport type amphibian was introduced late in 1927 and was a development that stemmed from the model OL-8 that was seeing so much useful service in the Naval fleet as an observation, spotting, and scouting airplane.

The very first in the long line of these unusual amphibious airplanes was in development for well over a year and was formally introduced to a handful of interested spectators in Jan. of 1925 at Bolling Field in Washington, D. C. In it's basic form, this prototype was a 2 place open cockpit biplane of rather large proportions with double-bay wing panels; introducing the unusual "slipper-type" hull or float. This "float" was incorporated into the fuselage structure and mounted a retractable, wheeled landing gear. This combination was a simple answer to a thorny problem and was immediately hailed as an important contribution to the science of aircraft design. It's powerplant was the newly developed inverted "Liberty 12" engine of some 400 h.p.; an engine modification which was under development for some time, with this installation in mind. As for the airplane, it was purposely designed to be used as an observation-airplane or utility-scout, by the Army Air Service. To say the least, it actually astonished many and drew much favorable comment for it's practical utility, and quite good performance; outflying many landplane types of this day! This design was developed by Grover Loening and as was typical of most every one of his designs; it was very efficient and extremely airworthy. Historical accounts relate and prove that these early type "Loening Amphibians" made many memorable and record-breaking flights to all parts of the globe; demonstrating to all their hardy spirit and capable versatility. Of these pioneering achievements, Grover Loening and his associates could be justly proud.

It remains as fact, that Grover Loening was an ingeneous designer, and one of our foremost aeronautical engineers. While many

Fig. 129. Loening OL-8 type in service with U. S. Marines, shown here over Nicaragua. The OL-8 served as basis for design of "Cabin Amphibian".

Fig. 130. A Loening "Air Yacht" of the early "twenties".

others would design airplanes that were at best just improvements or just modifications over existing designs, Loening's designs were always a new approach to a problem; though very unorthodox in many instances! To keep some semblance of production going, as a few other aircraft manufacturers were doing during the turn of the "twenties", Loening designed and built efficient hi-lift wings for the memorable "DeHaviland 4" mail plane. These "Loening wings" added 10 m.p.h. to the top speed of the "DH-4" and allowed it to land 10 m.p.h. slower with very much better aileron control throughout the entire speed range. These wing panels were of the double-bay type and used interplane bracing consisting of streamlined steel tubing, in place of the normally used wood struts. (Though a conclusion from hearsay and hangar-talk, it appears that Loening had many sets of these wings left unsold and used them on the "amphibian"!) In years just prior to the advent of the "amphibian", Loening was building the "Air Yacht" which was also quite an interesting design. It was a "flying boat" of the high wing monoplane type, as pictured here, powered with various engines including the 12 cylinder "Liberty"; the engine was mounted in "pusher" fashion atop the wing with seating up forward in the hull for five.

Leroy Grumman, a very imaginative and able designer-engineer, joined Grover Loening in 1925 as his assistant and with Loening was largely instrumental in the design and development of the model OL-8 for use by the Navy Fleet. Later he worked on the development of the passenger-carrying cabin amphibians. Some years later, Grumman became quite famous for his "fighter types" developed for the Navy

Fig. 131. "Liberty" powered Loening amphibian in service with the Air Corps, note unusual slipper-type float.

air-arm; notably the "FF-1", the "Wildcat", and the "Hellcat" series.

To continue discussion on the Loening Cabin Amphibian", that was built under this certificate number, it was a rather large biplane with the typical "Loening" double-bay wing panels. By deepening the fuselage shape a trifle, into a cabin section, ample room was provided for 4 to 6 passengers in enclosed comfort. The pilot sat up forward in an open cockpit, just behind the 9 cylinder Pratt & Whitney "Wasp" engine of 410 h.p. that swung a 3-bladed "prop"; the "3-blader" was calculated to keep down the over-all diameter and to provide sufficient clearance at the hull. The type certificate number for this model was issued in April of 1928 and it was proudly displayed at the Detroit Air Show of that same year. Upon completion of service tests, and with the suggestion of prospective buyers, it was decided that more power would be desirable for a craft of this type. The design was then modified somewhat to take the 9 cylinder Pratt & Whitney "Hornet" engine of 500 h.p. Needless to say, this made into a much better combination; for discussion on this more powerful version, see chapters for A.T.C. #66 and #67 in this volume. The "Loening Amphibians" were manufactured by the Loening Aeronautical Engineering Corp. in New York City. The roster at Loening Engineering was Grover Loening, President; Albert Loening, Vice President & Treasurer; Leroy Grumman, General Manager; and B. C. Boulton was Chief Engineer.

Listed below are specifications and performance data for the "Wasp" powered Loening "Cabin Amphibian"; span both 45'8", chord both 72", wing area 500 sq ft., airfoil "Loening", length overall 34'8", height wheels up 11'5", height wheels down 12'9", empty weight 3730, useful load 2170, payload 1150, gross weight as 7 place 5900 lb., max. speed 116, cruise 97, land 54, climb 770, ceiling 12,000 ft., gas cap. 140 gal., oil 12 gal., range 570 miles. Price at factory posted in May of 1928 was $24,700. The fuselage framework was built up of spruce longerons, uprights, and diagonals, with dural gussets at every joint; then it was covered with "Alclad" sheet. The wing panels were built up of laminated spruce spars and stamped out "Alclad" aluminum ribs, fabric covered. The landing gear retraction was manual and required about 12 seconds. Wheel brakes and a metal prop was provided, as was an electric inertia-type engine starter. Records checked do not reveal the number of these airplanes that were built, but from all indications we can assume that there was only a prototype built in this version.

A.T.C. #35
(4-28)
OX-5 INTERNATIONAL, F-17

Fig. 132. The OX-5 powered "International", model F-17.

The Model F-17 as pictured here in the various views, was an all-purpose airplane especially designed for the varied services performed by the average small operator. As was typical for a plane of this type, it was a 3 place open cockpit biplane and was powered with the 8 cylinder Curtiss OX-5 or OXX-6 engines. As all of the "Internationals", and this one was surely typical, it had the characteristic octagonal shaped wood framed and veneer covered fuselage that was a trade-mark for the type.

The distinctive "International" type were designed and developed by Edwin M. Fisk who plainly acknowledged that he got the idea for the "8 sided" fuselage configuration, used on all of the "Internationals", from that used in dirigible construction. Ed Fisk had been designing and building airplanes since some time back in 1910, from then until 1925 he had managed to design and build some 11 assorted models. In the next couple of years he had designed and built some 6 more! The history on all of these earlier models is buried deeply somewhere in the secrecy of the past, but the first "International" type as we came to know it, was built as a "Catron-Fisk" at Venice, Calif. late in 1924. "International", as a firm, was formed and started operations on Feb. 9th of 1927 by 3 men and a lot of ideas. They built some 32 airplanes in that year, most of these were powered with the OX-5 engine, but a few models were powered with the "Hisso". About a year later, the force of 3 men had grown to about 100, and they were building an average of 3 planes a week.

Speaking with a number of old-timers, one could assume that the "International" was quite impressive in it's behavior and often considered a little better than the average, and certainly at least on a par with the best; but they just didn't seem to possess the appeal of the "competition" in this class of airplane. The "F-17" was known and described as a "soft and smooth airplane", due to it's noise and vibration absorbing all-wood construction, no doubt. They were very stable and easy to fly, but probably not as described by one old-timer of many hours, who was prejudiced no doubt because of his love for the "International". He would say with sincerity. "they could just as well put in an easy-chair for the pilot". The F-17 was far from being a ravishing beauty, but it did impress one as a stalwart and capable airplane.

The International Aircraft Corp. seemed to have difficulty in staying financed and con-

Fig. 133. The "International" was often used by men of business to make calls quickly and cheaply.

sequently did a good bit of moving around. After their stay at Long Beach, Calif. they re-organized and moved to Ancor, Ohio for a while and then on to Jackson, Mich.; there they quietly folded up some time around 1930. This is not to be considered detrimental nor unusual, because there were many, many, aircraft companies, and some of long standing too, that had to fold up in 1930 or shortly thereafter. The "crash" of the market and the following "depression" was grimly felt by one and all!

The type certificate number for the OX-5 powered model F-17 was issued in April of 1928, and it was continuously built through 1929. For a later version of the OX-5 powered F-17, see chapter for Grp. 2 approval numbered 2-100 which will be discussed later. The Model F-17 was also offered with the "Hisso" and "J5 Whirlwind" engines, see chapters for A.T.C. #154 and #155.

Listed below are specifications and performance data for the OX-5 powered "International" model F-17; span upper and lower 35', chord both 60", wing area 326 sq. ft., airfoil U.S.A. 27, length 25', height 9'6", empty weight 1480, useful load 660, payload 320, gross weight 2140 lb., max. speed 98, cruise 82, land 35, climb 520, ceiling 10,500 ft., gas cap. 30 gal., oil 4 gal., range 325 miles. Price at Long Beach factory was $2750 less engine. The fuselage framework was built up of spruce members and plywood gussets and covered with mahagony veneer plywood. The wing panels were built up of spruce box type spars and spruce and plywood built-up ribs, the wing panels were covered with fabric. Many earlier "Fisk" type had used the "I strut" for interplane bracing but this model had the normal "N" type struts of streamlined steel tubing. The engine radiator was blended into the center-section of the top wing. The long exhaust pipes for the engine were buried in the cowling and had an outlet at the bottom, this was to keep exhaust gases well below the level of the cockpits. The landing gear was of fairly wide tread and used rubber in compression for shock absorbers.

Fig. 134. Note "8-sided" fuselage in this view of the "International".

A.T.C. #36
(4-28)
OX-5 PHEASANT, MODEL H-10

Fig. 135. 1928 OX-5 powered "Pheasant" model H-10.

Added to the steadily growing list of "OX-powered" airplanes was the "Pheasant" biplane; it was no great aeronautical wonder, but it certainly was an airplane of good behavior and very pleasant lines. This normally would have been almost a cinch combo for a fair acceptance among the flying-folk, but circumstance had dealt rather unkindly with this airplane in it's brief history, and it never did get a fair chance to make any great lasting impression on the people in the trade. The name got to be fairly well known around the country, but the airplane itself remained quite rare. More's the pity too, because it had interesting possibilities and would have been deserving to take it's place in the nation's budding aircraft industry. Though this was many years ago now, memory can still recall vividly how we were so thoroughly impressed with our first good look at a "Pheasant" biplane that was visiting the local airport. It's lines were catchy, and it's performance shown on the numerous flights that afternoon was impressive to say the least; pilots who were fortunate enough to get a try at it, were very enthusiastic.

As was typical of the trend and the times for airplanes leveled at the so-called commercial or small-operator market, the "Pheasant" was also a 3 place open cockpit biplane. It was very well proportioned with trim and simple lines and was powered with an OX-5 engine of 90 h.p. It had wings of unequal span that gave it a nice slant, there was dihedral on the lower panels only, with a normal amount of interplane stagger and an aileron in each wing. The engine was neatly cowled in with removable sections; the radiator was of the nose-type and was blended into the nose-section with numerous louvers provided for ample air-flow out of the engine compartment. The robust landing gear was of the long leg split-axle type and had a fairly wide tread, the tail-skid was of the spring-leaf type. The "Pheasant's" method of construction harbored no innovations and was typical of the period, it's performance was adjudged to be at least on a par with the best. Eleven "Pheasants" were reported built in 1927 and it's type certificate number was issued in April of 1928.

The "Pheasant" biplane was originally designed and developed in early 1927 by Lee R. Briggs, who operated the local airport and flying-school just out of Memphis, Mo. The airplane was in the course of being tested for approval and due to an unfortunate circumstance, Lee Briggs was killed in a crash during a test flight. S. J. Wittman then took over as test-pilot for further development. Early in 1928, the design rights and a major portion of the company's holdings were bought up by

Adolf Bechaud and associates, and the former company was reorganized into the Pheasant Aircraft Co. of Fond Du Lac, Wisc. Various other members of the new firm were Tom Meiklejohn Jr., Rowinski, Florian Manor, and the now celebrated S. J. "Steve" Wittman who was their test-pilot in charge of development.

Over-anxious and over-eager to get things under way, 3 airplanes were completely built before receiving the "Dept.'s" o.k., this was a grave misjudgement and lead to additional set-back. Meanwhile, the company was also carrying out much experimentation on various projects. This included the Pheasant "Traveler" which was a perky little one place light cabin monoplane with a full cantilever wing. The powerplant for this model was also a company development, it was a 4 cylinder vertical in-line air-cooled engine of 55 h.p. that was built around various components of the Ford 'Model A" engine. About 3 of the "Traveler" were built. They were shown at the Detroit Aircraft Show of 1929 and created a mild stir of interest, but planes of this type were not yet of general interest to the flying public. A considerable amount of money had been spent in the development of the "OX-5 Pheasant" and later on with the "Traveler", plus it's engine, and soon the companies financial position became somewhat precarious; especially since the infamous "Depression" was making itself felt by now. This had put a crimp in further development and production, so consequently the company was sold. The new owner never entered production with either model. The former members of the firm now went their separate ways; Rowinski was soon busy helping to design and develop the "Tank" engine which was basically an air-cooled OX-5. S.J. "Steve" Wittman as everybody knows, later became famous for his uncanny ability with light racing airplanes, both as designer-builder and pilot.

In the 1928 Air Derby from New York to Los Angeles, "Steve" Wittman flew an OX-5 powered "Pheasant" biplane, the one shown here, and was maintaining a good strong 3rd place until he was forced down over the desert with engine radiator trouble. Despite the delay involved, he still managed to finish in 12th place. Of the 38 entries that started, only 23 went the full distance! Shortly after, in a return race from Los Angeles to Cincinnati, Wittman and the "Pheasant" picked up a 4th place win. This performance spoke well for both the airplane and the pilot. Of the "OX-5 Pheasants" that might still be in existence, there is knowledge of one that is being rebuilt to fly again. Incidently, it is the one that was owned by "Steve" Wittman, and is pictured here in various views.

Fig. 136. The Pheasant "Traveller" with engine designed around Ford "Model A" components.

Listed below are specifications and performance data for the OX-5 powered "Pheasant" model H-10; span upper 32'4", span lower 29', chord upper 62", chord lower 60", wing area 283 sq. ft., airfoil Aeromarine 2A Mod., length 23'6", height 9', empty weight 1351, useful load 738, payload 340, gross weight 2089 lb., max. speed 100+, cruise 85, land 40, climb 550, ceiling 12,500 ft., gas cap. 38 gal., oil 4 gal., range 400 miles. Price at factory was first $2375, later raised to $2650, and was $2895 in 1929. The fuselage framework was built up of welded chrome-moly steel tubing, faired to shape with wood fairing strips and fabric covered. The wing panels were built up of solid spruce spars and wood built-up ribs, also fabric covered. The fabric covered tail-group was built up of welded chrome-moly steel tubing, the fin was ground adjustable and the horizontal stabilizer was adjustable in flight. Pontoons were available for $1,000 extra.

A.T.C. #37
(4-28)
TRAVEL AIR, MODEL 8000

Fig. 137. The Travel Air "8000" with 4 cyl. Fairchild-Caminez engine.

Although a very interesting airplane and well bally-hooed for a time, the "Travel Air" model 8000 was actually little known and soon forgotten, a very rare type and only a few were built. It was introduced in the latter part of 1927 with high hopes and great enthusiasm because it was to be the first "Travel Air" to be powered with a medium horsepower air-cooled radial engine of American manufacture, and it showed such a great promise of being a success; but "dame fortune" dealt rather harshly with it.

The model 8000 was basically a standard "Travel Air" biplane that was powered with the new 4 cylinder Fairchild-Caminez engine of 120 h.p. This rare "Travel Air" combination actually came about through the expectant, though short-lived popularity of the "Caminez" engine. Soon there were Caminez-powered versions of numerous other standard airplane makes. New models were excitedly promised by such firms as "Waco", "Kreider-Reisner", "Boeing", "Fairchild", and others, but as it turned out, most were just prototypes and remained a "one of a kind". The Model 8000 seemed to be the most successful of the lot and was the only "type" to be certificated with the Caminez engine.

The "Caminez" engine was designed by Harold Caminez, it provoked much interest and discussion about the country because it operated on such a novel principle. It used a "figure 8 cam" instead of the normal crankshaft to transmit power to the propeller drive. It's 4 short-stroke pistons were all linked together to transmit their power to the "cam", this method speeded up the normal "cycle" by two and produced a low and efficient r.p.m. at the "prop" during normal engine speeds. This principle was really quite novel and produced very efficient propulsion, but a certain amount of buyer's hesitance had developed and many "bugs" were still present in the engine that were never worked out properly. Consequently, the engine was finally discontinued, almost bearing the stigma of being a flop.

In it's basic form, as pictured here, the 8000 was a standard 3 place open cockpit biplane of the type currently built by Travel Air during this period. It is of interest to note the relatively high engine mounting that was necessary to provide ground clearance for the large diameter propeller, that is, without altering the landing gear. Kreider-Reisner solved this problem by using a 3-bladed or 4-bladed prop. The unusually large propeller diameter was necessary and made possible by the engine's low "prop r.p.m.", this actually delivered a more efficient propulsion and better performance than a small diametered high-speed prop.

The type certificate number for the "Travel Air" 8000 was issued in April of 1928 and the exact number of this type that were built is not known, only the evidence of a prototype. The Model 8000 was flown in the 1928 National Air Tour by J. Nelson Kelly and finished in 13th place, not a bad performance considering the field of tough competitors it had to contend with. The "8000" was manufactured by the Travel Air Mfg. Co. at Wichita, Kansas.

Listed below are specifications and performance data for the "Caminez" powered "Travel Air" model 8000; span upper 34'8", span lower 28'8", chord upper 66.75", chord lower 55.75", wing area 296 sq ft., airfoil "T.A. #1", length 24'6", height 8'11", empty weight 1475, useful load 825, payload 360, gross wt. 2300 lb., max. speed 110, cruise 92, land 42, climb 700, ceiling 12,000 ft., gas cap. 42 gal., range 500+ miles. Price at the factory was somewhere near $5,000. The fuselage framework was built up of welded chrome-moly steel tubing, the rear section was braced with steel tie-rods, it was faired to shape with wood fairing strips and fabric covered. The wing panels were built up of laminated spruce spars and spruce and plywood built-up ribs, also fabric covered. The interplane struts were of streamlined chrome-moly steel tubing and the interplane bracing was of streamlined steel wire. The fabric covered tail-group was built up of welded steel tubing, the fin was ground adjustable and the horizontal stabilizer was adjustable in flight.

The 4 cylinder "Caminez" engine was rated to 120 h.p. at 960 r.p.m., cruise r.p.m. was about 900 and the gas consumption at this r.p.m. was about 7.5 gallons per hour. At a peak r.p.m. of 1000, the power rating jumped to 135 h.p. The Caminez engine sold for $2480. Plagued with roughness and somewhat excessive oil consumption, the Caminez was also tried in an 8 cylinder version, but the results were not much better so it was shelved for possible later development. The type was never revived. The next development in the "Travel Air" biplane was the "Siemens" powered Model 9000, see chapter for A.T.C. #38 in this volume.

A.T.C. #38
(4-28)
TRAVEL AIR, MODEL 9000

Fig. 138. The "Travel Air" model 9000 was a pleasant engine-airplane combination.

The good looking "Siemens-Travel Air", or model 9000, was also a standard 3 place open cockpit biplane of the type currently built by Travel Air during this period, and it was typical in most all respects except for the powerplant installation. The engine in this case was the highly-rated 9 cylinder "Ryan-Siemens" of 125 h.p. Actually this was a "Siemens-Halske" engine of German design and manufacture. The "Travel Air" biplane and the "Ryan-Siemens" engine made into a really pleasant combination and a well proportioned thing of beauty, but it was always somewhat of a rare sight due to the small number built and never was afforded the chance to build up much of a following. It had a rather short run, and before long was discontinued.

A Model 9000 of this type was flown in the National Air Tour for 1928 by Geo. B. Peck and finished in 22nd place; had the 9000 been spared from it's many little misfortunes while on the tour, it would have no doubt, placed very much better. One "Siemens-Travel Air" did manage to come into a little bit of temporary prominence, when 17 year old Richard James flew "solo" across the country from New York to California. His aim was to win a $1000 prize that was put up by the "A.S.P.A." organization, the American Society for Promotion of Aviation. His "Travel Air" performed admirably and was appropriately named the "Spirit of American Youth". The type certificate number for the "Travel Air" model 9000 was issued in April of 1928 and it was built in small numbers by the Travel Air Mfg. Co. at Wichita, Kansas.

Listed below are specifications and performance data for the "Ryan-Siemens" powered "Travel Air" model 9000; span upper 34'8", span lower 28'8", chord upper 66.75", chord lower 55.75", wing area 296 sq. ft., airfoil "T.A. #1", length 24'4", height 9', empty wt. 1475, useful load 825, payload 360, gross wt. 2300 lb., max. speed 112, cruise 93, land 42, climb 700, ceiling 12,000 ft., gas cap. 42 gal., range 450 miles. Price at the factory was upwards of $5,000. Most all of the construction details were similiar to other "Travel Air" biplanes of this period, see previous chapter. The landing gear was of the normal split-axle type and used two "spools" of wound rubber shock-cord to snub and absorb the loads, wheel brakes were available. All

"Travel Air" biplanes were now sporting a streamlined head-rest for the pilot. The "Siemens" installation in the "9000" differed somewhat from the average because it did not use the typical front mounted collector-ring that was so strongly associated with the "Siemens" engine. The accompanying illustration shows its clean and neat appearance. The standard color scheme was "Travel Air Blue" for the fuselage and tail-group with silver wings, but some of this model were done up in custom colors. The "Ryan-Siemens" engine put out 125 h.p. at 1800 r.p.m. and cost $2970 less propeller.

One "Travel Air" biplane, quite similiar to the 9000 shown here, was built in a version that was powered with, as one who claimed that he was quite familiar with the species so aptly put it, that "awesome contraption". Referring to the 10 cylinder twin-row radial aircooled "Anzani" (French) engine of 120 h.p., of course. This ship was built and licensed under a Grp. 2 approval that was issued in July of 1928. For further discussion on this particular model, see chapter for Grp. 2 approval numbered 2-25 which will be discussed later. For some whimsical reason, this airplane was called "Smith's Incubator"!

The "Travel Air" model 7000 was quite a rare version of the biplane type and was listed as a 5 place combination cabin transport. It was powered with either the 180 "Hisso" engine or the "J4 Whirlwind" engine and could be said to resemble a Boeing 40-B-4 in it's cabin layout, although there actually was no 40-B-4 yet when the model 7000 was brought out. The first of this type was built as a special purpose airplane in 1926.

The 1928 Detroit Air Show held in April of that year had 69 airplanes on exhibit, and thousands upon thousands of interested visitors proved that the average public was surely becoming air-minded; the amount of sales rung up at the show kept the industry humming along for many months to come

For the next "Travel Air" development, which was the J5 powered monoplane Model 6000, see chapter for A.T.C. #100 in this volume. For the "Warner Scarab" powered biplane, the model W-4000, see chapter for A.T.C. #112.

A.T.C. #39
(5-28)
BERLINER "PARASOL", CM-4

Fig. 139. The handsome Berliner "Parasol" with Curtiss OX-5 engine.

Not exactly an oddity but most always the subject of mixed feelings and curious interest, the "Berliner" parasol-type monoplane was actually little known and was somewhat of a rare type. It was a pleasant looking and very interesting airplane but it's shining light seemed to have been held under a bushel! Through various circumstances, it was never afforded the proper chance to stimulate and win a general acceptance among those that would have need for an airplane of this type. The design of this airplane was a fairly new approach to getting "3 in the air with an OX-5" and though performance, reliability, and utility, was as good as or even better than the average, and pilot comments were always favorable and even enthusiastic, the average pilot-owner of these times, just couldn't seem to break away from the "old reliable two-winger"!

The "parasol" monoplane was actually a pleasant and efficient configuration that possessed an inherent "pendulum stability" and usually afforded a very good range of vision; a feature especially desirable while flying around close to home in heavy round-the-airport traffic. Oddly enough, the "Berliner Parasol" was the first and only OX-5 powered "parasol" monoplane to be certificated, although there were a number of other "parasols" built that never went beyond the experimental or X-stage. An OX-5 powered "Berliner" monoplane was flown in the Class A division of the Transcontinental Air Derby for 1928 from New York to Los Angeles by Sam Turner and finished in 15th place.

As pictured here, the model CM-4 was a 3 place open cockpit monoplane of the parasol type with a strut braced semi-cantilever wing of thick section placed well above the fuselage for convenience and better visibility. The "Berliner" was designed to take advantage of the efficient monoplane configuration but still adhered to the traditional open cockpits! The model CM-4 was the OX-5 powered version of this airplane; it was a sturdy, stable ship of very good performance and handled exceptionally well. The type certificate number for the CM-4 was issued in May of 1928 and from all indications present, only a mere handful were built. Two other versions of this model were also being offered, the model CM-5, which was almost identical except that it was powered with a Wright J5 engine, it's performance with this amount of power should have been just short of phenomonal. The other model, the CM-6, was also typical except that this version was powered with the hardy little 7 cylinder Warner "Scarab" engine of 110 h.p. The "J5" and "Scarab" powered versions of the "Berliner" parasol monoplane were indeed a rare type

Fig. 140. The B/J "all vision" monoplane of 1929-1930. This was the firm's last commercial development.

and known records do not relate the number that were built, only the evidence of a porot-type in the CM-5 and CM-6. They may have been just "drawing board" models; although the only changes necessary to this airplane would have all occured ahead of the firewall. There was some opinion here and there that the "OX-5 Berliner" monoplane would have sold well had it been given more of a chance to prove it's worth, but when the original company, headed by Henry A. Berliner was reorganized, they elected to drop it from production and go on to bigger things; bigger things which unfortunately never did materi-alize, at least not in the commercial aircraft field.

These "parasol monoplane" beauties were designed and developed by Henry A. Berliner, a modest and versatile engineer who also designed and developed the interesting "Ber-liner Helicopter" that made a number of suc-cessful though very short flights before Army Air Service observers in 1924. The first of the "Berliner" monoplanes discussed here were built by the Berliner Aircraft Co. at Alexandria, Va., and then reorganized into the Berliner-Joyce Aircraft Co. Temple Joyce, the other half of the firm, was formerly an Air Service development test-pilot and was also formerly with "Curtiss" and "Chance Vought", his various abilities were well known in the industry. The "B/J" company later went into further reorganization and moved to a new plant and larger quarters in Baltimore (Dun-dalk), Md., with William Wait Jr. as their chief engineer. The new firm also planned bigger things, first coming out with an "all vision" high wing cabin monoplane that was indeed an interesting type, as shown here; great possibilities were present in this design but it never went beyond the prototype test stage. Various other types were experimented with, but nothing outstanding was developed until the advent of the Berliner-Joyce two-seated "Pursuit Ship" of the early "thirties". This was the Air Corp's P-16, which was a Curtiss "Conqueror" powered gull-wing bi-plane that seated a pilot and a gunner in tan-dem open cockpits. This was a "type" of very good performance, but it's success was short-lived and sometime after, the company was dissolved and completely absorbed into the North American Aviation Corp.

Listed below are specifications and per-formance data for the OX-5 powered "Ber-liner" parasol monoplane model CM-4; wing span 36', chord 75", wing area 219 sq. ft., length 25'10", height 8', empty weight 1490, useful load 810, payload 380, gross wt. 2300 lb., max. speed 105, cruise 90, land 41, climb 470, ceiling 12,000 ft., gas cap. 40 gal., oil 4 gal., range 470 miles. Price at the factory was $3290. The fuselage framework was built up of welded chrome-moly steel tubing, heavily faired to shape with wood fairing strips and fabric covered. The sturdy semi-cantilever wing was built up of spruce box-type spars and spruce and plywood built-up ribs, also fabric covered. The outer wing struts were of chrome-moly steel tubes encased in balsa-wood fairings, the struts of the center-section cabane were of streamlined steel tubing. The landing gear was of the normal split axle type and the engine radiator was mounted below the engine compartment as a free-air type. The "all vision" B/J monoplane as shown here was also tested with a 4 cylinder in-line "Cirrus" engine of 86 h.p.

A.T.C. #40
(5-28)
CURTISS, "OX-5 ROBIN"

Fig. 141. 1928 "OX-5 Robin" as built by Curtiss.

Through the years up to this point, Curtiss had ventured only occasionally into the building of commercial airplanes, with types like the early "Oriole", the "Lark", and the "Carrier Pidgeon" to name a few. To Curtiss these were mostly a sideline because they were mainly interested in and pretty well content to specialize in the Army and Navy types which managed to keep them quite busy. By keeping well informed of the trends in the industry, it was getting to be the general belief at Curtiss, in these days, that the open cockpit biplane though far from being obsolete yet, had probably run it's course and would soon be outmoded for general purpose use or for light commercial work. It was their belief that the light cabin monoplane would be much more suited for this type of use, due to it's offering of performance advantages and greater comfort and utility. With this in mind, the engineering team at Curtiss proceeded to design and develop a light cabin monoplane which was to become the 3 place "Robin".

As everyone knows, all of the "Curtiss" airplanes were appropriately named after birds, and thoughtfully enough they always picked a "bird" that was more or less in character with the personality of the airplane. The most famous of these during this period were the "Falcon", the "Condor", and the fabulous Curtiss "Hawks". Thus, the "Robin" came along to take it's place among this family of birds!

Much research and development testing went into the "Robin" so consequently it was a well thought out airplane; it wasn't a radical type by any means, but an incorporation of many of the latest design features and performance requirements deemed desirable for a private-owner type aircraft. A three place airplane was still considered the ideal, so the "Robin" seated 3 in a fully enclosed cabin that provided very good vision with ample room and comfort. There was a sky-light for overhead vision and full length side-windows at the pilot's station, a useful feature that was popularized by the Fairchild and the Ryan. The pilot sat in a bucket-type seat up front and the two passengers sat side by side on a bench-type seat in back. On this early model there were 3 doors for easy entrance for each occupant, on later production models the left hand rear door was eliminated. In it's basic form, the "Robin" was a strut braced semi-cantilever high wing monoplane of rather "boxy" but very pleasant lines, and surprisingly so, was aerodynamically efficient. Curtiss proved it by wind-tunnel tests before the airplane was ever built! This first model was powered with the Curtiss OX-5 engine, of which Curtiss still had a good supply on hand. The engine was cowled in neatly and the coolant radiator was blended into the nose-section; a wood "prop" was standard and the use of a "spinner" was optional. The robust landing gear was of the wide tread outrigger type that was tied into the wing strut truss, using "Rusco" rubber discs enclosed in a streamlined casing for absorbing the bumps; the tail-skid was also "Rusco" sprung and swiveled with the rudder, this in itself was a good economical substitute for the help of wheel brakes.

The original "Robin" in it's "X" version, as shown here, had "airfoiled struts" and also used a strut junction truss typical to that used on the Bellanca; this was to stiffen the wing at this point and to eliminate the possibility of wing deflection during heavy aileron loads. In a later modification into the "C" or licensed version, also pictured here, the "airfoiled" struts were replaced with struts of streamlined steel tubing and as can be seen, all junctions were carefully streamlined; the engine cowling was amply louvered for better cooling. This was the "speed job" version of the "Robin" and it proved it's mettle well in numerous "OX-5 races" about the country. "Casey" Jones flew it at the Air Races of 1928 held in Los Angeles. Earlier in the year this "Robin", still as the "X" version, was flown in the 1928 National Air Tour by Dan Robertson and came in just about dead-last (23rd)!

This particular airplane as shown in it's "X" version and in it's "C" version, was the prototype and the first of the very popular and even quite famous Curtiss "Robin" series, and it was also the first Curtiss commercial type airplane to be certificated. The Curtiss Aeroplane & Motor Co. at Garden City, L. I. had worked out the design and the development of this model and after building a few more for test, turned over all manufacturing chores to the Curtiss-Robertson Aircraft Co., a subsidiary, whose plant was built at St. Louis (Anglum), Mo. in about March of 1928. Production was started on the "Robin" in about July of that year. See chapter for A.T.C. #68 in this volume for accounts of the production version of the "OX-5 Robin". The original Curtiss "Robin" type was first introduced along about March of 1928 and it's type certificate number was issued in May of that year. The "Robins" that were built under this A.T.C. were built by Curtiss, the parent company, and various reports show that about 4 were built for test purposes. For the "Curtiss" development of the "Challenger Robin", see chapter for A.T.C. #63 in this volume.

Listed below are specifications and performance data for the OX-5 powered Curtiss "Robin"; wing span 41', chord 72", wing area 224 sq. ft., airfoil "Curtiss C-72", length 25'10", height 7'10", area of airfoiled struts 41 sq. ft., empty weight 1480, useful load 737, payload (with 30 gal. gas) 387, gross weight 2217 lb., max. speed 99, cruise 84, land 44, climb 450, ceiling 10,200 ft., gas cap. normal 30 gal., gas cap. maximum 50 gal., range 380-590 miles. The fuselage framework was built up of welded chrome-moly steel tubing, lightly faired to shape and fabric covered. The wing was built up of routed spruce spars and "Alclad" aluminum stamped out ribs, also fabric covered. There was a 25 gallon fuel tank in each wing root for a maximum capacity of 50 gal. A wood "prop" and metal "spinner" were standard, but the Curtiss-Reed "bent slab" metal propeller was optional. Standard color scheme was burnt-orange fuselage, yellow wings, and black trim.

Fig. 142. The "Robin" was gentle and had good flight characteristics.

A.T.C. #41
(6-28)
WHIRLWIND-WACO, ASO

Fig. 143. 1927 "J5 Waco" flown to 9th place in 1927 Ford Air Tour by Charlie Meyers.

The ever-popular "Whirlwind-Waco" was an airplane of appealing personality and in it's brief time of actual manufacture, managed to acquire a staunch legion of enthusiastic admirers; and numerous names and designations. But in spite of all the various names and designations used for this airplane, it was still basically typical to the standard "Model 10" in most all respects except for the engine installation and some modifications made necessary for this combination. The powerplant for this model was the 9 cylinder Wright "Whirlwind J5" engine of 220 h.p.; more often than not this combination was quite fondly refered to as the "Whirlwind-Waco", the "J5 Waco", or later on as the "J5 Straight-Wing". As pictured here in the various views, in very good likeness, it was a 3 place open cockpit biplane. A trim and rather handsome airplane of simple lines.

Like a dormant talent waiting to be discovered, the added power of a "J5" gave this "Waco" a brilliant and responsive performance that belied her otherwise docile behavior. With power to spare, she became a lively and graceful beauty in the air, a ship that was well adapted to any man's taste for aerobatics; which is but a polite name for the every-day brand of "stunt flying". To this capability most anyone would heartily agree to, after seeing some "hot rock" pilot put her through the many hair-raising manuvers she was so deftly capable of. Some certain manuvers, like "rolls" for instance, took a little doing on the part of the pilot, but the results looked good and were quite satisfying.

Besides...this airplane was surely no slouch for everyday cross-country flying either, getting in and out of all sorts of airfields and landing strips, under all manner of conditions. This "knack" or ability was many times proven in the various "air tours" and "air derbies" held about the country. Four "J5" powered "Waco Sports" were flown in 1927 Ford Air Tour and came in to 5th, 7th, 9th, and 12th place. These being flown by John P. Wood, John Paul Riddle, Chas. W. Meyers, and E. W. "Pop" Cleveland respectively. Later in 1927 a "J5 Waco" was flown by John P. Wood in the Air Derby from New York to Spokane to a 4th place behind two "hot shot" Lairds and a warmed-over Buhl "Airster". This adept ability to overcome rigorous conditions, almost inevitable in this type of flying, was displayed again when John P. Wood flew a "Whirlwind-Waco" to first place in the 1928 National Air Tour; yet further proved by Charlie Meyers when he brought his "Whirlwind-Waco" in to 4th place. The reason "air tours" were mentioned here with some bit of emphasis, is because they were "reliability tours" and better proved an

Fig. 144. The original J5 installation, it takes but little imagination to hear those "short-stacks" popping!

Fig. 145. Charlie Meyers, chief pilot at "Waco". Shown with his mascot "Whirlwind Jimmie" during 1927 Ford Air Tour.

airplane's adaptability to varied conditions around and about the country. The "tour" was a far better criterion of an airplane's all-round efficiency than the more specialized events like air races "around a pylon", etc.

The "Whirlwind-Waco" was introduced in 1927 as the "Ten-W" (10-W) or "Waco Sport", and about 19 were built in that year. During the course of it's eventful existence, this airplane managed to acquire numerous names and factory designations. Starting out as the "Ten-W" or "Waco Sport", it was also known as the "J5 Waco", the "Whirlwind-Waco", or the "J5 Straight-Wing" to distinguish it from the "J5 Taper-Wing". It was also known as the "220" and finally as the model "ASO". Among some of the better known enthusiastic users and owners of the "J5 Waco", and they were many, was E. W. "Pop" Cleveland. "Pop" Cleveland was that amiable flying-salesman for "Aerol Struts" who used his plane to make his country-wide sales promotion rounds; his "Waco Sport" was one of the first of this model to be built. The "Whirlwind-Waco" appealed to so very many, but naturally there were those that had not the price for a spanking new "J5 Waco"; so, some enterprising fellows have been known to dig up a good used "J4" somewhere and work it into the old "Waco 10". From then on, the forsaken OX-5 was "tarp" covered, and lay to gather dust in a far corner of the hangar or shed.

The type certificate number for the "Whirlwind-Waco" was issued in June of 1928 and this model was in continuous manufacture until sometime in 1929, when the trusty J5 was replaced by the new 7 cyl. "Whirlwind" of the "J6 series" that was rated at 225 h.p. For a discussion on this model, the CSO, see chapter for A.T.C. #240 which will be discussed later. The "Waco" was manufactured by the Advance Aircraft Co. at Troy, Ohio. Clayton J. Bruckner was Pres., Charlie Van Sicklen was Mgr. for sales, and Chas. W. Meyers was chief pilot in charge of design and development.

Listed below are specifications and performance data for the "Whirlwind J5" powered "Waco" model ASO; span upper 30'7", span lower 29'5", chord both 62.5", wing area 288 sq. ft., airfoil Aeromarine Mod., length 22'6", height 9'2", wheel tread 78", empty wt. 1550, useful load 1050, payload 490, gross weight 2600 lb., max. speed 125, cruise 105, land 44, climb 1050, ceiling 19,000 ft., gas cap. 65 gal., range 575 miles. Price at the factory in 1928 was $7215, in early 1929 the price had gone up to $7335. The following are weights and performance figures for the earlier "light job" of 1927; empty weight 1411, useful load 899, payload 360-470, gross weight 2310 lb., max. speed 126, cruise 108, land 40, climb 1200, ceiling 19,500 ft., gas cap. 37-54 gal., range approx. 400-500 miles. The "Whirlwind-Waco" could be operated with either wheels, skis, or floats; metal propeller and wheel brakes were standard equipment. Wings were wired for lights. The construction of this model was more or less typical to the other "Waco" types, except that the J5 installation added greatly to the performance and consequently imposed greater stresses, especially on the wing structure. These stresses were absorbed and overcome by "beefing up" the wing spars and using heavier gauge material for interplane struts and wing-wire bracing.

A.T.C. #42
(6-28)
HISSO-WACO, DSO

Fig. 146. The 180 "Hisso" and the "Waco" made into a hairy-chested combination.

Every once in a while you will hear an airplane described as "hairy-chested", this would also apply to the "Hisso-Waco". The "Hisso-Waco", or "Hisso Straight-Wing", was also known as the model "150" and "180" and later on as the DSO. It was basically typical to the standard "Waco 10" in most all respects, it's major differences were the modifications that were necessary for the installation of the "Hisso" (Hispano-Suiza) engines. Either model of these engines was optional, the low-compression "Model A" of 150 h.p., or the hi-compression "Model E" of 180 h.p. Actually, there were other models available of the so-called "Hisso" engines, but these two versions were the most popular in this country. These engines were of the water-cooled "vee type", having 8 cylinders in two "banks" of 4 each, and using an overhead cam-shaft for each bank of cylinders. For a while, they were also easily available as war-surplus stock and at a fairly reasonable price too, but were somewhat more complex and a trifle harder to repair and maintain. In a comparison with the Curtiss OX-5 engine, they were not very economical and under pasture-airport conditions were generally costing a good bit more to operate and maintain.

In it's basic form as shown here, the "Hisso-Waco" was a 3 place open cockpit biplane with a rather lively and very satisfying performance, but with these engines it seemed to have a somewhat limited appeal and not very many of this model were built, possibly about 45 were built altogether. The "Waco 10" with "Hisso" engine was first introduced in the latter part of 1927 and only one was built in that year, the type certificate number for the "Waco" with "Hisso" was issued in June of 1928.

At least one example of these venerable "Hisso-Waco" type are known to be still flying in practically it's original form, as shown here. Naturally, it causes much comment of all sorts every time it flies. In spite of the sometimes tiresome task of "propping" the engine by hand, and taxiing without the help and convenience of brakes and a tail-wheel; most of the "old airplane" enthusiasts tend to be purists in this respect and would prefer to have their airplanes authentic to the period and "type", and not modified nor modernized in any way! The "Hisso-Waco" had a very good reputation as all "Wacos" did, naturally, but it's number was spread pretty thin about the country, and for most people it remained a rare sight. Manufactured by the Advance Aircraft Co. at Troy, Ohio.

Listed below are specifications and performance data for the "Hisso" powered "Waco" model DSO; span upper 30'7", span lower 29'5", chord both 62.5", wing area 288 sq. ft., airfoil Aeromarine Mod., length 23'6", height 9'2", empty wt. 1508, useful load 896, payload 410, gross wt. 2404 lb., max. speed 115-120, cruise 100-105, land 44, climb 700-750, ceiling 17,000-18,000 ft., gas cap. 37 gal., range 375 miles. The double figures listed above are for the model "150" and "180" respectively. A later version, the model DSO, was typical except for the following figures; empty wt. 1600, useful load 1000, payload 440, gross wt. 2600 lb., max. speed 120, cruise 105, land 47, climb 700, ceiling 17,000 ft., gas cap. 65 gal., range 600 miles. Price at the factory in 1929 was $3935 for the "Hisso A", and $4085 for the "Hisso E".

The fuselage framework was built up of welded chrome-moly steel tubing, faired to shape with wood fairing strips and fabric covered. The wing panels were built up of heavy spruce spars and wood built-up ribs, also fabric covered. The interplane struts were of streamlined chrome-moly steel tubing and the interplane bracing was of streamlined steel wire. Like the "J5", the "Hisso" installation added greatly to the performance and consequently imposed greater stresses, especially on the wing structure. These stresses were absorbed and overcome by "beefing up" the wing spars and using heavier gauge material for the interplane struts and wing-wire bracing. The fabric covered tail-group was also built up of welded chrome-moly steel tubing, the fin was ground adjustable and the horizontal stabilizer was adjustable in flight. The rugged landing gear was of the familiar "oleo spring" type, using 30 x 5 wheels, brakes were available. The fixed tail-skid was of the practically trouble-free "spring leaf" type and mounted a replaceable hardened shoe. An inertia type engine starter and a metal propeller were available as optional equipment. The standard color scheme was the familiar two-toned fuselage in "Waco Green" with silver wings, later some were available in blue and silver and in maroon and silver. The last of the "Hisso-Waco" type was built in the latter part of 1929. For a discussion about the fabulous "Taper-Wing", see chapter for A.T.C. #123 which will be discussed later. For the BSO with the 5 cyl. Wright engine, see chapter for A.T.C. #168.

Fig. 147. One of the last of the DSO type, rebuilt by Denny Trone of California.

A.T.C. #43
(6-28)
SIMPLEX "RED ARROW", K2S

Fig. 148. Simplex "Red Arrow" K2C, powered with 90 h.p. Kinner engine.

The trim and saucy little Simplex "Red Arrow" monoplanes, though fairly well known to everyone in the flying fraternity about the country, were somewhat of a rare type because they never were built in any great number. This was a very interesting type, at least it was a departure from the general run of private-owner type airplanes that were being offered to the flying public up to this time. Actually, the "Red Arrow" was more of a high performance sport-type airplane and well capable of performing the complete retinue included in the art of aerobatics of this period. A point in fact that was often and well proven by the antics of H. S. "Dick" Myhres during his frequent demonstrations about the country; and many, many, times over the homefield in Defiance. Stories say, that he used to "scare the pants off people" with his reckless type flying and devil-may-care abandon. On the other hand, the "Red Arrow" could be docile and well-mannered and quite well suited as a training ship; being able to handle equally well the chores of both primary and secondary pilot-training.

In it's basic form, as shown here, the "Red Arrow" was a roubst little strut braced semi-cantilever mid-wing monoplane that seated two side by side in an open cockpit. The powerplant in this model, the K2S, was the new 5 cylinder "Kinner" engine of 90 h.p.; this was a combination that offered enough extra horsepower for a seemingly effortless and quite snappy performance. H.S. Myhres flew a Kinner powered Simplex "Red Arrow" to first place in the Class A "speed dash" from San Francisco to Los Angeles, this was during the 1928 National Air Races held at Mines Field. He also won a 2nd, 3rd, and 4th place in several other free-for-alls, pitted against many capable pilots and many fine airplanes, some of much greater horsepower. This display of performance spoke well for the "Red Arrow" and the deft capabilities of one "Dick" Myhres.

Earlier versions of the Simplex "Red Arrow" were powered with the early "75 horse" Kinner engine which had A.T.C. #3 in the engine category and was one of the first air-cooled "radial" engines designed especially for the light private-owner type airplanes that were beginning to show up on the commercial-aircraft market. The type certificate number for the Simplex model K2S was issued in June of 1928 and it was manufactured by the Simplex Aircraft Corp. at Defiance, Ohio.

This was the first mid-wing monoplane to be certificated and though it was a popular layout amongst some of the builders of light and ultra-light airplanes, the wing placement was surely controversial to say the least.

The mid-wing never proved to be very popular nor too desirable on the general market for civilian type airplanes. Yet, despite it's various limitations, real or imaginary, the "mid-wing" did have some qualities still attractive to many experimenters and light plane home-builders throughout all these years and even to the present.

The Simplex Aircraft Corp. was formed early in 1928 with E. J. Allen as it's Pres., Geo. H. Roberts as Sec-Treas., and their designing engineer was O. L. Woodson. Woodson was an old-timer of broad experience and great ability, and is probably best remembered for the development of the "Woodson" line of commercial biplanes; notably for the development of the "Woodson Express", a "Hisso" powered cargo and mail carrier of 1926. Later on, the chief pilot in charge of test and development, and other shenanigans, for the Simplex Co. was the inimitable H. S. "Dick" Myhres who flew the "Red Arrow" monoplanes with such carefree and reckless abandon. All the while proving their spirit and deft capabilities. Simplex Aircraft was always an enterprise of quite modest proportions, with a force at most of some 25 or 30 craftsman of all sorts, and they had ample facilities for their modest production in a plant well situated next to a rail-siding. Later on they had promoted an adjoining air-field.

Listed below are specifications and performance data for the Kinner powered Simplex "Red Arrow" monoplane model K2S; wing span 34'4", chord 60", wing area 150 sq. ft., airfoil Clark Y, length 22'3", height 7', empty wt. 1020, useful load 572, payload 190, gross wt. 1592 lb., max. speed 120, cruise 108, land 37, climb 1000, ceiling 15,000 ft., gas cap. 35 gal., oil 5 gal., range 550 miles. Price at factory started out at $3995., later raised to $4115., and finally to $4500. A later version of the K2S was more or less identical except for the following figures; empty wt. 1100, useful load 685, payload 223, gross wt. 1785 lb., gas cap. 40 gal., range 650 miles. Performance remained very much the same due to the installation of the improved Kinner K-5 engine which was rated at 100 h.p. The fuselage framework was built up of welded chrome-moly steel tubing, faired to shape with wood fairing strips and fabric covered. The wing framework was built up of spruce box-type spars with spruce and plywood built-up ribs, also fabric covered. The unusual wing bracing was a mass of struts going in every direction! This is not true...but they really were a complex system.

The Simplex "Red Arrow" model K2S was the first airplane to be certificated with the "Kinner" engine, a 5 cylinder air-cooled "radial" engine that was rated at some 75 h.p. when it was first introduced late in 1927 by "Bert" Kinner. The "Kinner" was awarded one of the first A.T.C. numbers (#3) in the engine category, with a rating of 90 h.p. at 1810 r.p.m. All of the early "Kinners" had uncovered valve-action. The well known Kinner K-5 was brought out late in 1928 with a rating of 100 h.p., and the valve-action was now enclosed in rocker-boxes. The K-5 weighed in at about 250 lb. and the fuel consumption at cruise r.p.m. was 6 gallons per hour. For the model K2C (a "cabin" version of the K2S), see chapter for A.T.C. #44 in this volume.

Fig. 149. Simplex "Red Arrow" monoplane in prototype version, powered with 75 h.p. Kinner engine.

A.T.C. #44
(6-28)
SIMPLEX "RED ARROW", K2C

Fig. 150. A rare photo of the Simplex "Red Arrow" Coupe. This prototype was the K2C version.

Anyone who can remember seeing the Simplex "Red Arrow" monoplane model K2C, is indeed a rare individual, because this type must have been literally as scarce as "hen's teeth". There is only one picture to show of this model. The K2C version of the "Red Arrow" monoplane was more or less identical to the K2S previously described, except for a "canopy" over the 2 place open cockpit; this made it into a cabin plane. Performance, in comparing the two models, was said to be slightly improved for the "cabin model", but recorded performance figures showed it as being the same, and basic specifications were identical.

The mid-wing configuration in reference to the cockpit placement, had an inherent tendency to blank off a direct view downward, so the "Red Arrows" had little side-windows in the lower level of the fuselage at the occupants feet for visibility in that direction. Visibility above and in turns was excellent. Another interesting and characteristic feature of the Simplex monoplanes was the unusual strut arrangement of the wing bracing truss. On the later models, this was somewhat improved by addition of a stub-wing. The type certificate number for the model K2C was issued in June of 1928 and it was also built by the Simplex Aircraft Corp. at Defiance, Ohio.

For the next development in the "Red Arrow" series, see chapter for A.T.C. #238 which will follow later; this chapter discusses the Simplex model W2S that was powered with the Warner "Scarab" engine. The "Red Arrow" models remained more or less the same clear into the early "thirties", but never did achieve any great measure of popularity. Sad, but very true, the mid-wing monoplane was sort of shied away from by the flying public during this early period of airplane development, and the low-winged monoplane was said to be absolutely poison! It is true that our knowledge of aerodynamic behavior of the "mid-wing" and "low-wing" was meagre, but there were many wild stories in circulation about these two types and this was harder to combat than the aerodynamic problems.

A word or two about O. L. Woodson, designing engineer at Simplex Aircraft would be quite appropriate here. Woodson was a modest man of versatile background, sound ability, and ample talent; he previously developed many interesting models in the "Woodson" biplane series. Notable among these was the "Hisso" powered "Woodson Express" that was designed for contemplated use on the short feeder-lines proposed to flow into the transcontinental air-mail system during 1926. Unfortunately, no great demand developed for these and the biplane series were discontinued. In 1927, Woodson designed and developed the model "M-6", a two-seated low-winged light monoplane that was powered with the 5 cylinder Detroit "Air Cat" engine; this model showed a great promise, but financial difficulties and what have you, nipped that project practically in the bud. The "Simplex Red Arrow" type was more or less a development from the "Woodson" model M-6.

Listed below are specifications and performance data for the "Kinner" powered Simplex "Red Arrow" monoplane model K2C; wing span 34'4", chord 60", wing area 150 sq. ft., airfoil Clark Y, length 22'3", height 7', empty wt. 1070, useful load 570, payload 190, gross wt. 1640 lb., max. speed 120, cruise 108, land 37, climb 1000, ceiling 15,000 ft., gas cap. 35 gal., oil 5 gal., range 550 miles. Price at factory was $4495 and later raised to $4995. A later version of the K2C was identical except for the following figures; empty wt. 1150, useful load 648, payload 193, gross wt. 1798 lb., gas cap. 40 gal., range approx. 650 miles. The fuselage framework was built up of welded chrome-moly steel tubing, faired to shape with wood fairing strips and fabric covered. The wing framework was built up of spruce box-type spars and spruce and plywood built-up ribs, also fabric covered. The fabric covered tail-group was built up of welded steel tubing. Construction of the "canopy" on this model is doubtful, but typical to the average it would more than likely be a steel tube frame covered with metal-edged pyralin.

A.T.C. #45
(6-28)
WILLIAMS, "TEXAS-TEMPLE"

Fig. 151. A good likeness of the rare "Texas-Temple" monoplane.

This little-known airplane, in development since 1926, was a "rare bird" indeed and according to the somewhat elusive and sketchy information available on this type, possibly no more than three of these airplanes were ever built. Of a fairly plain and normal configuration, it was offered either as a single place or 3 place open cockpit semi-cantilever monoplane of the "parasol type". The wing was placed rather low over the fuselage to afford the pilot better vision, pilot could see both above and below the wing with only a very small arc of blind-spot. Designed primarily for high performance, it was powered with the 9 cylinder Wright J5 of 220 h.p. Two versions of this basic type were planned for production. The single place model had a hatch covered compartment up forward of the pilot's cockpit for stowing mail and express cargo, of which it could carry a payload of some 500 lbs. Performance characteristics of this version were quite good and it was therefore suitable and also offered as a single place hi-performance "sport model" for the sportsman-pilot. This single place model was called the "Speedwing". The 3 place version of this airplane seated all 3 occupants in open cockpits, the passenger's cockpit was up forward under the wing, with the pilot occupying the rear cockpit, to put him in position for better visibility. This 3 place model was called the "Commercial-Wing". Being basically typical, the performance and flight characteristics of these two versions were more or less similiar.

As pictured here, the generally trim and basically simple appearance of this airplane was somewhat reminiscent of the "Ryan" M-1 type or probably more like the "Yackey Sport" parasol monoplane. The "Texas-Temple" had been under development since 1926, probably with 10 cyl. "Anzani", and it's type certificate number was issued in June of 1928. As of July in 1928, two were built in the 3 place version, and one was built as a single place cargo-carrier. The production schedule was tentatively set at one completed airplane every 8 weeks.

The "Texas-Temple" was designed and developed by George W. Williams, a pioneer in aircraft design and a capable air-man of Temple, Texas. Williams reportedly built his first airplane, a light monoplane, some 50 or more years ago in 1908, and thus launched the activities of the Texas Aero Mfg. Co. This enterprise was the first, and for a long time the only "aeroplane factory" in the state of Texas. To Geo. Williams, with the cooperation of George Carroll, also goes reported credit of developing the first successfull,

Fig. 152. 1929 Texas-Temple "Sport", modified with installation of Cirrus engine. Experimental version, not produced in quantity.

full monococque, full cantilever wing. This was indeed a revolutionary aeronautical development, but according to recorded data it was never put to use. Constantly seeking improvement in his designs, George Williams tested his airplanes continuously; Williams was finally killed in a crash while test-flying one of his own airplanes. As far as can be determined from available data, the company folded up shortly after and further development of the "Texas-Temple", and cabin-job built along these lines, was dropped. The "Texas-Temple" parasol monoplane was manufactured at rather modest quarters at Temple, Texas by the Texas Aero Mfg. Co., it was also listed for a time as the Geo. Williams Airplane Mfg. Co. Later in 1929, the remnant of the firm was reorganized into the Texas Aero Co. of Dallas, Texas. They planned production in 1930, but records are hazy of their activities. A Texas "Sport" was built with "Cirrus" engine.

Listed below are specifications and performance data for the J5 powered "Texas-Temple" three place Commercial-Wing"; wing span 42', chord 72", wing area 235 sq. ft., airfoil U.S.A. 27, length 25'10", height 8'3", empty wt. 1350, useful load 950, payload 395, gross wt. 2300, max. speed 130, cruise 112, land 35, climb 1500, ceiling 18,000 ft., gas cap. 65 gal., oil 8 gal., range 650 miles. Price at factory was $9500-$10,000. For the single place "Speed-Wing" version, all specifications were the same except for the following figures; empty wt. 1350, useful load 1055, payload 500, gross wt. 2405 lb., max. speed 135, cruise 118, land 42, climb 1450, ceiling 17,500 ft., fuel and oil capacity were same, price at the factory was $9500. Specs and data for the prototype model were the same except for the following figures; wing span 39'4", chord 72", wing area 225 sq. ft., empty wt. 1160, useful load 1090, payload 567, gross wt. 2250 lbs., max. speed 138, cruise 120, land 35, climb 1500, ceiling 18,500 ft., gas cap. 50 gal., oil 8 gal., range 600 miles. The prototype was tried earlier with an OX-5 and "Hisso" engines.

The method of construction for this airplane was simple and quite conventional. The fuselage framework was built up of welded chrome-moly steel tubing, the portion back of the pilot's cockpit was braced with steel tie-rods, the framework was lightly faired to shape and fabric covered. The engine mount was detachable by 4 bolts, allowing varied engine isntallations. The semi-cantilever wing was built up of solid spruce spars and gusseted spruce built-up ribs, fabric covered. The wing was built up as a continuous unit but the spars were joined in the center by a splice; the 50-65 gallon fuel tank was placed in the center portion of the wing. There was a semi-circular cut-out in the trailing edge of the wing, for better vision upward and for better access to the pilot's cockpit. Wing struts were of chrome-moly steel tubing, encased by wooden ribs of airfoiled section and fabric covered; the struts on some of the ships were simply encased in balsa-wood fairings and fabric covered. Ailerons were of welded steel tube construction and fabric covered. The wing panel was fastened to the fuselage in parasol fashion at the center-section by 4 inverted "vee" struts that straddled the front cockpit. The long-leg landing gear was of fairly wide tread (96"), and rubber compression rings were used as shock absorbers. The fabric covered tail-group was built up of welded chrome-moly steel tubing, the fin was ground adjustable and the horizontal stabilizer was adjustable in flight. Stabilizer adjustment was by crank, giving a fine adjustment to an infinite number of position settings. The average stabilizer adjustment was by lever and plate with possibly 8 or 10 notches for various settings, this arrangement was unhandy for a setting just in between. The "Texas-Temple" offered wheel brakes, inertia type engine starter, and metal propeller as optional equipment. Wings were wired for lights.

A.T.C. #46
(6-28)
BUHL "SPORT AIRSEDAN", CA-3C

Fig. 153. The high performance of the "Sport Airsedan" was leveled at the sportsman-pilot.

The "Sport Airsedan" by Buhl was a de-luxe 3 place cabin sesqui-plane of rather high performance and various custom features, but in it's basic form, it was still quite similar to the standard type of "Airsedan". Being strictly a high performance sport model, especially leveled at the sportsman-pilot, the model CA-3C had a somewhat limited appeal in general and therefore not very many were built. This version of the "Airsedan" now had lower wings that were tapered in plan-form and section, and were very small in comparison to the upper wing; in fact, about one quarter of the area, making it a true so-called "sesqui-plane" or "sesqui-wing" type. All of the Buhl "Airsedan" series being built after this period were of the sesqui-wing cellule arrangement that was used by Buhl to such good advantage for a very efficient configuration and an exceptionally strong wing truss.

The model CA-3C as pictured here in the various views, was powered with a 9 cylinder "Whirlwind J5" engine of 220 h.p. One of this type was flown by Louis Meister, their chief pilot since 1926 and now promoted to Sales Mgr., in the National Air Tour for 1928 and finished it in 10th place among some very tough competition. Lee Schoenhair and "Tom" Colby flew a "Sport Airsedan" to 6th place in the 1928 Air Derby from New York to Los Angeles, right on the heels of a bunch of "Waco" Taper-Wings and Laird "Speed-Wings". The type certificate number for the "Sport Airsedan" model CA-3C was issued in June of 1928 and it was built into 1929 by the Buhl Aircraft Co. in their plant at Marysville, Michigan. The price at the factory field for this model in 1928 and part of 1929 was $11,000, and though a trifle high, it was still considered a good buy for this type of ship. Later in the year of 1929, the "Sport Airsedan" was also offered in a version that was powered with the new 9 cylinder "Whirlwind" engine of the improved "J6 series" that was rated at 300 h.p. For a discussion of this model, the CA-3D, see chapter for A.T.C. #163 that will follow later. A "Sport Airsedan" that was built in 1928, bearing a serial #28, and having a registration number of NC-5860, was reported still flying somewhere in the mid-west, and almost in it's original condition, some 28 years after it was built!

Listed below are specifications and performance data for the "Whirlwind J5" powered Buhl "Sport Airsedan" model CA-3C; span upper 36', span lower 20'10", chord upper 72", chord lower (tapered) 35" M.A.C., wing area 240 sq. ft., length 28', height 8', empty wt. 1760, useful load 1440, payload 660, gross wt. 3200 lb., max. speed 134, cruise 112,

Fig. 154. View showing sesqui-wing arrangement of "Sport Airsedan".

land 47, climb 800, ceiling 16,000 ft., gas cap. 90 gal., oil 5 gal., range 840 miles. Price at factory field was $11,000. The fuselage framework was built up of welded chrome-moly steel tubing, lightly faired to shape with wood fairing strips and fabric covered. The wing framework was built up of spruce box-type spars and spruce built-up ribs, also fabric covered. The gasoline supply of 90 gallons was carried in two tanks, one in each upper wing root. The landing gear was of the split-axle type, but differed from that used on the standard "Airsedan" in that it did not fasten to the lower wing. Wheel brakes, inertia-type engine starter, and a metal propeller were standard equipment.

Discussing the various "Buhl" airplanes with an old-timer one day, it was agreed upon and later verified, that there was also a model CA-3B, which was known as the "Junior Airsedan". It was a pert 3 place cabin job that was also of the sesqui-plane type and a good deal similiar to the "Sport Airsedan" except that it was quite a bit smaller and was powered with a 7 cylinder Warner "Scarab" engine of 110 h.p. Four were reported built in 1928. It is just a guess of course, but it would seem that Buhl miscalculated on this one. It would seem that this is the model that really had possibilities for private-owner acceptance, but Buhl elected to drop it's manufacture and no doubt missed a good bet. But one never knows!

Fig. 155. Louis Meister, on right, flew this "Sport Airsedan" to 10th place in 1928 National Air Tour.

A.T.C. #47
(6-28)
BELLANCA, CH-200

Fig. 156. The Bellanca CH-200 with Wright J5 engine.

"Bellanca" has long been a proud and dominant name in the annals of aviation history, and these few words will hardly be sufficient to do it any fair justice. The history of "Bellanca" has been a brilliant one, and the name has been almost synonomous with airplane efficiency. We might say, the first "Bellanca" airplane of any note was the 6 place "Wright-Bellanca", the model WB-1, that is pictured here in one of the illustrations. It was a high wing cabin monoplane of inspiring and very advanced design, a ship that was formally introduced at the New York Air Races in October of 1925. It's main duty was that of promotion and being a test-bed for further development of the newly introduced Wright "Whirlwind" J4 engine, a 9 cylinder air-cooled radial engine of 200 h.p. The introduction of this engine by Wright was very timely and it soon became very popular and very much in demand; even for commercial installations. The combination of the WB-1 with the "Whirlwind J4" engine was such a departure from the average configuration and so advanced in design, that more than one expert was prone to be skeptical of it and stated somewhat emphatically that many of it's features, tho' highly desirable, must surely yet be proven in actual service!

The model WB-2, a modified and improved version of the "Wright-Bellanca", was introduced later in 1926 and is also pictured here. This new ship incorporated quite a few worthwhile modifications in it's structure and was powered with the newest "Whirlwind" engine of the "J5 series". The J5 was also very much improved over the J4 and was rated at 220 h.p. Together, the new "Bellanca" and the new "Whirlwind" made quite a combination. The "Wright-Bellanca" WB-2, by the way, was the airplane that Chas. A. Lindbergh tried to buy for his New York to Paris flight, but both Wright and G. M. Bellanca were rather reluctant to sell and they couldn't agree to any sort of deal that would be satisfactory to all parties concerned. Ironically enough, this was the same airplane that Chamberlin and Levine flew to Germany about a month or so after Lindbergh's successful flight to Paris in a "Ryan" monoplane! Both of these "Wright-Bellancas", the WB-1 and the WB-2, ran rough-shod over competition throughout the years of 1925 and 1926, winning most all of the efficiency races and setting an example for performance that was hard to beat. These two fine ships were the prototypes of the typical "Bellanca" design; a design that became so familiar, even influenced aircraft design in general, and had endured practically without change for so many years yet to come.

Earlier "Bellanca" history can take us

Fig. 157. The Bellanca CH-200 was one of the most efficient airplanes of it's time.

Fig. 158. 1925 Wright-Bellanca WB-1 a revolutionary design, far ahead of it's time.

back a few years to the year of 1923. Guiseppe Mario Bellanca, as many other early pioneers, including Grover Loening and Igor Sikorsky, was engaged in building hi-lift and hi-performance replacement wing panels to earn a few "bucks" and to keep some semblance of production going, while working on new developments. These "wing sets" were built for the famous mail-carrying DH-4, the beloved "Jenny", and for just about anyone who cared to improve the performance of their airplanes with a good set of wings. G. M. Bellanca understood air-flow and became known as a "wizard with the airfoil"; he was using this skill and knowledge to some good advantage to keep the wolf away from his door during these lean and hungry years of aviation history.

Among some of the assorted models developed and built by Bellanca in 1923 and 1924, was a high wing cabin monoplane that carried 5 people with the dubious power of a 10 cylinder "Anzani" (French) engine rated somewhere between 90-100 h.p. Mind you, this was actually true, but it can be unbelieveable. This model, the "CF" Air Sedan, is pictured here. Other models available, were a six-seater powered with an OX-5 engine, and an 8-seater that was powered with the "Hisso A" engine of 150 h.p.

Early in 1927, Bellanca and the Wright Co. had a parting of the ways, and once more G. M. B. decided to go on his own. The former WB-2 was now known as the "Columbia", and was flown by Clarence Chamberlin and Bert Acosta to a new endurance flight record of over 51 hours without refueling, in Jan. of 1927. This was a preparatory test for the planned flight across the Atlantic Ocean. In June of 1927, with Chamberlin and Chas. Levine aboard, this same ship (C-237) winged it's way across the Atlantic to Eisleben, Germany after a flight of some 3900 miles. Later on in that year, a Bellanca "J" (NX-3789) was built for another assault on the endurance record, this ship became the "Pathfinder" in 1929 and was flown by Yancey and Williams to Rome, Italy from Old Orchard, Maine. Other "Bellancas" were setting records too; too many to mention!

The model CH was basically a passenger-carrying version of the Model J, and was introduced early in 1928. The Bellanca Model "CH-200", as pictured here, was G. M. Bellanca's first airplane to be certificated, the ticket was issued in June of 1928. In it's basic form, the CH-200 was a 6 place high wing cabin monoplane and was powered with the 9 cylinder Wright J5 engine of 220 h.p. Though somewhat of a refinement over the earlier models, it was still typical in many ways; Bellanca knew by instinct and proof that he had a very good basic design and felt there would surely be no point in changing it just yet. The CH-200 was a strut braced semi-cantilever high wing monoplane with a rather peculiar cantilever landing gear, that for obvious reasons earned it the name of "bow-legged Bellanca"! This model had proved itself fairly rugged and quite dependable, and it's performance was well above average. Late in 1928, Victor Dalin flew a CH-200 in the air races at Los Angeles and copped both the Aviation Town & Country Trophy and the Detroit News Air Transport Trophy for efficiency, winning over many entries that represented the cream of the industry. The first "Bellanca" plant was in Arlington, Staten Island, New York. Early in 1928, operations of the Bellanca Aircraft Corp. were moved to a new plant in New Castle, Delaware.

Listed below are specifications and performance data for the "Whirlwind J5" powered Bellanca model CH-200; wing span 46'4", chord 79", wing area 273 sq. ft., airfoil "Bellanca", length 27'9", height 8'6", empty wt. 2190, useful load 1860, payload 1120, gross wt. 4050 lb., (earlier models of the CH were much lighter), max. speed 126, cruise 106, land 46, climb 850, ceiling 13,000 ft., gas cap. 80-90 gal., oil 6 gal., range 700-800 miles. Price at factory was $14,050. The fuselage framework was built up of welded chrome-moly steel tubing, faired to shape and fabric covered. The wing framework was built up of routed spruce spars of "I beam" section and spruce and plywood built-up ribs, also fabric covered. The distinctive wing struts were built up of a framework of chrome-moly steel tubing employing ribs spaced across the strut to maintain the airfoil shape and then were fabric covered. The small strut cabane just inboard of the wing-attach point were designed to stiffen this portion of the strut juncture against excessive aileron loads. The "airfoiled struts" added 47 sq. ft. of lifting surface at an extreme dihedral angle which contributed to the airplane's stability a great deal. The "bow-legged" cantilever landing gear was of chrome-moly steel construction, and oddly enough used rubber shock-cord wound around pegs to absorb landing shock and loads. Two wing tanks of 40-45 gallons each were mounted one in each wing root for gravity fuel feed. Cabin interior was tasteful and quite spacious, with provisions for cabin heating. Soundproofing was used on later models.

The next development in this series was the "CH-300", which also was a "Bow-legged Bellanca", but was powered with the new Wright "Whirlwind" J6 engine of 300 h.p. This model was first built under a Grp. 2 approval, this approval was later superceded by A.T.C. #129.

Fig. 159. 1923 Bellanca CF, which could carry 5 people on the sputterings of a 10 cyl. Anzani engine.

Fig. 160. 1926 Wright-Bellanca WB-2, later flown by Clarence Chamberlin to Eisleben, Germany.

A.T.C. #48
(8-28)
STINSON "JUNIOR", SM-2

Fig. 161. The Stinson "Junior" SM-2 offered excellent performance on 110 h.p.

The model SM-2 as pictured here, was the baby of the Stinson monoplane line, and the very first of the popular Stinson "Junior" type; being formally introduced in the early part of 1928. In it's basic form, the "Junior" was a 3-4 place strut braced high wing cabin monoplane and typically a Stinson type in a scaled down fashion. The powerplant installation in this first model was the recently introduced 7 cylinder Warner "Scarab" engine of 110 h.p. The performance of this "baby Stinson" in this combination was quite excellent for a ship of this size. It was originally Ed Stinson's intent to offer this light cabin monoplane as a personal-type airplane for the private owner or the small business man, but the "Junior", also called the "Detroiter Jr.", was not exactly a dainty little sprite to begin with, and successive power increases in the interests of better and better performance caused it to become bigger and heavier. It was not too long before this "baby" of the Stinson line grew right out of the light-plane class.

Two SM-2 "Juniors" with Warner "Scarab" engines, were flown in the 1928 National Air Tour; one by Randolph Page, a former air-mail pilot, and one by Bruce Braun. Both of these men were excellent "Stinson" pilots, and finished the tour in 3rd and 6th place respectively. This was certainly a good showing for this airplane, considering the caliber of competition that was pitted against it. "Eddie" Stinson, flying a J5 powered "Detroiter" of the SM-1D type, finished in 5th place. Ed Stinson quite often imbibed in good natured braggin', and the performance of his airplanes in the 1928 "Tour", upheld his right to do so in very good style. In the 1928 Air Derby from New York to Los Angeles, an SM-2 flown by Geo. W. Hopkins made the 2939 mile trip in 34.5 hours of flying time; finishing in 11th place, (Class A).

This early version of the "Junior", with Warner "Scarab" engine, was not exactly a rare type but they were spread pretty thin around the country; about 27 were reported built. The type certificate number for this model was issued in August of 1928 and they were manufactured by the Stinson Aircraft Co. at Northville, Mich. One "Junior" of this

Fig. 162. 1928 Stinson "Junior" SM-2 with Warner "Scarab" engine, shown here during 1928 National Air Tour.

Fig. 163. The "Junior" was a typical Stinson in scaled-down fashion.

Fig. 164. Excellent example of restoration on a Stinson SM-2 of 1928 vintage.

early type was built into a test-bed for the 5 cylinder "Kinner" engine, for discussion on this model see chapter for Grp. 2 approval numbered 2-136 which will follow later. For the SM-2AA version of the "Junior" that was powered with the 5 cylinder Wright J6 engine of 150-165 h.p., see chapter for A.T.C. #145 which will also follow. The later built, improved type "Juniors", were fitted with more powerful engines and eventually evolved, through many modifications, into the very handsome and very popular "SR" or "Reliant" series.

Though a "Travel Air" biplane was the first airplane to mount the "Scarab" engine, the Stinson "Junior" was the first airplane to be certificated with this powerplant. Perhaps it would be fitting right here, to relate some background on this newly introduced engine. Warner Aircraft was organized in Oct. of 1926 for the production of "Scarab" engines. W. O. Warner, with full realization of the drastic need for low powered air-cooled "radials", developed a 5 cylinder and also a 7 cylinder type. The first 7 cyl. "Scarab 110" completed it's block tests in July of 1927, and then it was given a 150 hour flight test by Walter Carr, who had mounted the engine in his own "Travel Air" biplane. So, the "Travel Air" was a test-bed for the "Scarab" engine, but it was not certificated with this powerplant until much later; see chapter for A.T.C. #112 (W-4000).

Listed below are specifications and performance data for the Warner "Scarab" powered Stinson "Junior" model SM-2; wing span 41'5", chord 75", wing area 236 sq. ft., airfoil Clark Y, length 26'3", height 7'4", wheel tread 110", empty wt. 1516, useful load 984, payload 510, gross wt. 2500 lb., max. speed 106, cruise 90, land 42, climb 520, ceiling 10,000 ft., gas cap. 42 gal., oil 4.5 gal., range 425 miles. The fuselage framework was built up of welded chrome-moly steel tubing, faired to shape with plywood formers, spruce fairing strips, and fabric covered. The wing panel was built up of box-type spruce spars and spruce and plywood built-up ribs, also fabric covered. The air-foiled wing struts, designed to increase the plane's effective lifting area, were of chrome-moly steel tubing with wood ribs of "Clark Y" section or formed balsa-wood fairings, also fabric covered. The fabric covered tail surfaces were built up of welded chrome-moly steel tubing, the fin was ground adjustable and the horizontal stabilizer was adjustable in flight. The sturdy "oleo" type landing gear was of out-rigger form and of very wide tread, the tail-wheel was 11 x 3 and swiveled for better control while taxiing. With the tail-wheel mounted on the extreme end of the fuselage, the rudder was cut out to suit. The fuel supply was carried in two wing tanks, one on each side of fuselage, of 21 gallons each. An inertia-type engine starter, metal propeller, and individual wheel brakes were available. A propeller "spinner" was used on the early models, but the later models had a rounded nose-section and the prop-spinner was not used. A number of color combinations were available, but the most popular colors were black fuselage with orange wings, blue fuselage with cream colored wings, or a deep maroon fuselage with an orange-yellow wing. All colors were trimmed to suit, "Berryloid" finishes were used throughout.

A.T.C. #49
(7-28)
LOCKHEED "VEGA", MODEL 1

Fig. 165. 1928 Lockheed "Vega" with J5 engine.

The Lockheed "Vega" became a magic name in aviation circles because it was the "cleanest" and most advanced commercial airplane of it's time. Horsepower for horsepower it could out-fly them all! It was generously endowed with outstanding performance and a pure beauty that was balanced in perfect symmetry. A most unusual and very efficient design, the credit for which must surely go to John K. Northrop; a pioneer and a professed self-made aeronautical engineer who probably gave the air industry more "firsts" than any other man. Though we must be sure to mention that he was ably assisted on the "Vega" development by "Gerry" Vultee, who also became a great name in the air industry.

The very first of these fabulous "Vegas" was test-flown on the 4th of July in 1927, being especially built for Geo. Hearst of the San Francisco "Examiner", and was dubbed the "Golden Eagle". This bright-orange beauty is shown here in one of the views. After the necessary testing and preparation, it was entered and flown in the famous but somewhat ill-fated "Dole Derby" across the Pacific Ocean from Oakland, Calif. to Hawaii in August of 1927. On the crew were "Jack" Frost as pilot, and Gordon Scott as navigator and radio-man. After many hours out, it was presumed lost somewhere out in the broad expanse of the Pacific Ocean and was never heard from again, nor a trace of it ever found! It's ironic that this airplane should have vanished so completely without trace, because it was equipped with every known safety device to cope with most any emergency.

Another "Vega" pictured here (X-3903), was the third one built, built especially for the Wilkins Expedition to the Arctic Circle, (the second "Vega" had been hurriedly built as a company demonstrator and test-ship); the Wilkins "Vega" was first test-flown in Jan. of 1928 and was then flown by Carl "Ben" Eielson and Geo. Hubert Wilkins in the Arctic regions on the Wilkins Arctic Expedition of April in 1928. Making a 2200 mile exploration flight from Point Barrow to Green Harbor in Spitzbergen; over 1300 miles of this flight was over area never before seen by man! To his first "Vega", which he had called the "Los Angeles", he added another "J5 Vega", the "San Francisco", and both were used on the Hearst-Wilkins Anarctic Expedition of 1928-29. There making many flights in the South Pole regions while equipped with wheels, skis, or floats. Most of these exploration flights took place through the months of Nov., Dec., and Jan. In Feb. of 1929, they

Fig. 166. First Lockheed "Vega", the "Golden Eagle", was lost in the Pacific on a flight to Hawaii.

were both stored in a shed and later that year in Nov., the two "Vegas" were tuned up and prepared for another series of explorations in the Anarctic regions.

"Bob" Cantwell flew a J5 powered "Vega" in the 1928 National Air Tour and finished in 11th place amongst a terrific field of competition. Later in May of 1929, "Herb" Fahy in a "Whirlwind" powered "Vega" had set a non-refueling solo endurance record of 37 hours; the remarkable part of this flight was the fact that Fahy took off with 420 gallons of gasoline for a total gross weight of 5880 lb., this was surely a weight-lifting record for a plane of 41 foot wing span and a measly 275 sq. ft. of wing area! Even at that observers claim that he only used 2230 ft. of the run-way on take-off. In view of the foregone, even a partial list of the many other "Vegas" that pioneered in aviation history by setting records of all sorts would go on and on, so let it be sufficient to say that the Lockheed "Vega" type have earned themselves many brilliant accounts as consistent record-breakers.

A brief resume of early "Lockheed" history would go as follows. Allan and Malcom Loughead built their first plane in June of 1913, it was the "Model G" seaplane and it had flown continuously until 1918. Northrop was first with Loughead brothers in 1916 at Santa Barbara, Calif. There Northrop designed the F-1, a 10 place flying-boat of huge proportions. After W. W. I he designed the S-1, a folding-wing sport plane of 25 h.p., not blessed with any business to speak of, this enterprise folded in 1921. John K. Northrop joined Douglas Aircraft in 1923 and helped out on the design of the "World Cruisers". Lockheed Aircraft of Hollywood, Calif. was formed in 1926 to build the "Vega" and naturally, Northrop was with Lockheed again. Northrop left again, later in 1928 to study all-metal construction and flying-wing design, Gerry Vultee had now taken over as chief engineer.

By way of description, the Lockheed "Vega" Model 1, was a 5 place high wing cabin monoplane of all-wood construction. It's cigar-shaped fuselage was of the full monococque type, being of a laminated plywood shell construction, and the one piece wing was of the full cantilever type. This was a configuration that held parasitic resistance to a bare minimum and was to become the basic layout for all of the "Vegas" that followed. The powerplant for this model was the Wright "Whirlwind" engine of 220 h.p. which pulled the "Vega" around at a surprising speed of 138 miles per hour! The type certificate number for this "Vega" was issued in July of 1928 and it was manufactured by the Lockheed Aircraft Co. at Los Angeles, Calif., which was formed by the Loughead brothers, Allan and Malcom. The name was changed to "Lockheed" to avoid mispronunciation. John K. Northrop was Chief Engr. with "Gerry" Vultee assisting. For the "Wasp" powered version of the Lockheed "Vega", see chapter for A.T.C. #93 in this volume.

Listed below are specifications and performance data for the "Whirlwind J5" powered Lockheed "Vega"; wing span 41' chord at root 102", chord at tip 63", wing area 275 sq. ft., airfoil at root "Clark Y-18", airfoil at tip "Clark Y-9.5", length 27'8", height 8'6", empty wt. 1875, useful load 1595, payload 820, gross wt. 3470, max. speed 138,

Fig. 167. This "Vega" flew over both Arctic and Anarctic regions with Wilkins Expedition.

cruise 118, land 49, climb 850, ceiling 15,000 ft., gas cap. 96 gal., range 900 miles. Price at factory in 1928 was $12,500., later in the year it was listed as $13,500, in May of 1929 this was lowered to $13,000. The fuselage framework was built up of two plywood shells that were formed in a "concrete tub" and then assembled over a few circular wood formers that were held in line by a few wood stringers; after all necessary cut-outs were made, the fuselage was covered with fabric. The wing framework was built up of solid spruce spars and spruce and plywood built-up ribs, upon completion, it was then covered with plywood veneer and a fabric covering on top of that. The tail-group was of all-wood construction and covered in the same manner. It had been reported that some 52 "Vegas" with the "Whirlwind J5" engine had been built through 1929.

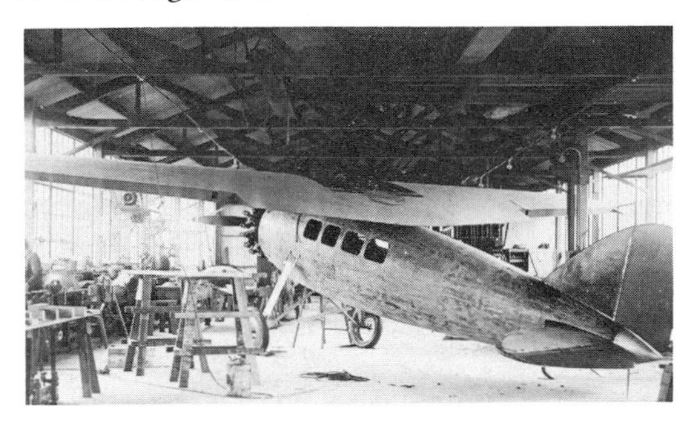

Fig. 168. "Golden Eagle", Vega No. 1, under construction in Hollywood, Calif. in early 1927.

A.T.C. #50
(7-28)
THE "HISSO-SWALLOW"

Fig. 169. 1928 "Hisso-Swallow"; the "Hisso" and the "Swallow" were made for each other!

The popular "Swallow" biplane when powered with the 8 cylinder vee-type "Hisso" (Hispano-Suiza) engines made into a mighty fine performing combination, but unfortunately, the war-surplus "Hisso" never did enjoy any real large amount of acceptance throughout the country; so consequently this model of the "Swallow" had a more or less limited appeal and only a small number of this version were built. Then too, most of these were custom built to order as special purpose airplanes, fit to the customer's various needs and demands.

In it's basic form, this model was typical to most any other "Swallow" coming off the line during this period, except for minor modifications necessary with the use of this engine installation. It was a 3 place open cockpit biplane and was most often powered with the "Hisso" Model A engine of 150 h.p., although it was also available with the hi-compression Model E engine of 180 h.p. The rugged character of the "Swallow" biplane was a natural for the "Hisso" engine, the added power boosted performance and general utility considerably and the installation blended in very nicely into the ship's fine classic lines. Pictured here are several splendid views of the "Swallow" with the "Hisso" installation, the pleasing and well balanced lines of this combination are readily apparent.

Three "Hisso-Swallows" were reported built in 1927, and the type certificate number for these models was issued in July of 1928; a few more were built in that year and then it was discontinued from production, although they were built on order. This version of the "Swallow" was manufactured by the Swallow Airplane Co. at Wichita, Kansas. Victor Roos, a familiar figure in aviation in and around Wichita of this period, was the V.P. and Gen. Mgr.

Listed below are specifications and performance data for the "Hisso" powered "Swallow" biplane; span upper 32'8", span lower 32'3", chord both 60", wing area 300 sq. ft., airfoil U.S.A. 27, length 24'3", height 9', empty wt. 1670, useful load 930, payload 400-500, gross wt. 2600 lb., max. speed 115, cruise 99, land 48, climb 780, ceiling 14,000 ft., gas cap. 40-60 gal., oil 5 gal., range 400-550 miles. The above figures are for the Model A version, the Model E version was identical except for the following figures; empty wt. 1728, useful load 972, payload 450-530, gross wt. 2700 lb., performance figures were proportionately higher. The fuselage framework was built up of welded chrome-moly steel tubing, faired to shape with wood fairing strips and fabric covered. The wing framework was built up of solid spruce spars and spruce and plywood built-up ribs, also fabric covered. All interplane struts were of chrome-moly steel tubing in streamlined section, interplane struts were of chrome-moly steel tubing in streamlined section, interplane bracing was of streamlined steel wire. The fabric covered tail-group was also built up of welded chrome-moly steel tubing, the fin was ground adjustable and the horizontal stabilizer was adjustable in flight. Wheel brakes and metal propeller were available. Price at factory averaged around $3750 with the "Hisso" Model A engine, the "E" version" was slightly higher. The next development in this series was the "J5 Swallow", for discussion on this model, see chapter for A.T.C. #51 in this volume.

A.T.C. #51
(7-28)
THE "J5 SWALLOW"

Fig. 170. The "J5 Swallow" was a handsome airplane, well suited for the sportsman-pilot.

The handsome and dashing "J5 Swallow" was a rather rare type in the "Swallow" line and it was somewhat of a pity that not very many of these were built, because it was a fine ship. An airplane of high performance and a good deal of glamourous quality. Among the small number of enthusiastic owners and boosters of the "J5 Swallow" was "Hoot" Gibson, the once very famous cowboy movie star that enjoyed it's spirited nature immensely and was rightfully very proud of it. The The "Whirlwind-Swallow" was generally quite typical to the other "Swallows" in production at this time, except for the engine installation which in this case was the 9 cylinder Wright J5 engine of 220 h.p., and other modifications made necessary for this combination.

In it's basic form, the "J5 Swallow" was a 3 place open cockpit biplane with a few deluxe features not normally found in the other standard models. Features such as metal propeller, individual wheel brakes, extra instruments, extra fuel capacity for a greater operating range, and a custom finishing job throughout, which was usually tailored to the customer's liking. We feel unfortunate that it had never been our good fortune to see a "J5 Swallow" put through the paces, but from accounts related by others in days gone by, we can easily assume that though in spite of it being strong-willed; when thoroughly mastered, it was more than able to hold it's own with the very best of them, and a pure joy to fly.

The "Whirlwind-Swallow" was somewhat of a pioneer too, in it's earlier days. Varney Air Lines, one of our pioneer airmail-carriers who operated between Salt Lake City and Pascoe, Wash. on the rugged C.A.M. #5 route, had a fleet of five "Swallows" that they used on this feeder-route in the northwest in 1926. Their ships were the newly developed "Swallow Mailplane" of 1925, which had been modified somewhat from the standard model in production at that time. At first these airplanes were powered with the 6 cylinder inline Curtiss C-6 engine of 160 h.p., but this combination lacked the necessary power to provide an adequate performance reserve for flying over the treacherous mountainous territory, so they were soon after re-fitted with the 9 cylinder Wright "Whirlwind" J4 engine of 200 h.p.; and from then on, everything worked out real fine. This early mail-carrying "Swallow" is pictured here in one of the views. Walter Varney's line was the first private carrier in the contract air mail system (C.A.M.) and after a number of mergers and re-organizations, became known as the origin of the mammoth "United Air Lines System".

A J5 powered "Swallow" of the type pictured here, was flown in the 1928 National Air Tour by Milton Aavang and finished in 15th place amongst a stellar field of 28 airplanes that were competing all-out for the coveted trophy. The comely Ruth Elder, who flew across the Atlantic as co-pilot with George Haldeman in the Stinson "Detroiter" named the "American Girl"; was a woman-pilot of ample ability and a great "Swallow" fan. She flew a "Whirlwind-Swallow" in the Women's Air Derby from Santa Monica, Calif. to Cleveland, Ohio in August of 1929, finishing in 5th place. Refer to chapter for A.T.C. #16 in this volume for a more detailed account of Ruth Elder's unsuccessful ocean jaunt. The type certificate number for the "J5 Swallow" was issued in July of 1928 and it was manufactured by the Swallow Airplane Co. at Wichita, Kansas. Victor Roos was V.P. and Gen. Mgr. for a time. Quite a few airplane companies had their plants in Wichita by now, and it was being proudly hailed as the aircraft manufacturing center of the U.S.A.

Listed below are specifications and performance data for the "Whirlwind J5" powered "Swallow" biplane; span upper 32'8", span lower 32'3", chord both 60", wing area 300 sq. ft., airfoil U.S.A. 27, length 23'6", height 9', empty wt. 1716, useful load 984, payload 450, gross wt. 2700 lb., max. speed 128, cruise 109, land 45, climb 1000, ceiling 18,000 ft., gas cap. 60 gal., oil 5 gal., range 550 miles. Price at factory in early 1929 was $8500, in 1928 the price was somewhat higher. The fuselage framework was built up of welded chrome-moly steel tubing, faired to shape with wood fairing strips and fabric covered. In fact, all construction details for the "J5 Swallow" were more or less typical to other standard "Swallow" models, see previous chapter. Extra instruments were usually provided, a metal propeller and wheel brakes were standard equipment. Inertia-type engine starter was available and the wings were wired for lights. Color schemes usually varied, according to customer order, but a combination that was quite popular was a jet-black fuselage with bright orange-yellow wings. For the "Swallow" biplane that was powered with the 7 cylinder "Axelson" engine (F28-AX), see chapter for A.T.C. #125.

Fig. 171. The "J4 Swallow" of 1926, Varney Air Lines began service with five of these.

A.T.C. #52
(7-28)
FOKKER "SUPER UNIVERSAL"

Fig. 172. 1929 "Super Universal" served routes in the Pacific south-west with Standard Air Lines.

The hard-working "Super Universal" by Fokker was a direct development and a logical modification of the earlier "Standard Universal", although it was a somewhat larger airplane and it was fitted with some 200 extra horsepower to handle the larger payload and provide the necessary increase in performance that was being demanded by operators all over the continent. By using a full cantilever wing construction, it eliminated the need for the familiar external wing bracing struts that were typical of the "Universal" type. This form of bracing was still used on the first few of the prototype "Wasp-Universal". The landing gear was modified and simplified to some extent and the cabin section was now fully enclosed, even for the pilot. In it's basic form, the "Super" was an 8 place high wing cabin monoplane that was powered with a 9 cylinder Pratt & Whitney "Wasp" engine of 400-420 h.p., easy removal of passenger seats allowed for a 1250 lb payload of freight. In it's manner of construction and assembly, it was of the typical and traditional "Fokker" type; having a welded steel tube fuselage framework and an all-wood cantilever wing.

The "Super Universal" was designed to work and was readily available with wheels, skis, or floats, and being a ship of very good performance in any combination; it was considered quite versatile and therefore very popular with the so-called professional "bush pilots" of the north country, especially in Canada and Alaska. It was also an ideal size and a money-maker too, for the small air-line and was used on a number of feeder-routes; it was used on scheduled and chartered trips, and also for the transportation of business executives. For years they performed many interesting and valuable services, and some are still giving efficient and reliable service in many far away corners of the globe, even to this day!

The accompanying illustrations show the "Super Universal" in it's natural habitat as fitted with wheels, skis, or on pontoons; equally at ease in any combination and doing all sorts of jobs, anywhere, at any time. Two "Super Universals" of the early type, were built late in 1927 and they are listed in the chapter for Grp. 2 approval numbered 2-3 which will be discussed later. The later modified version of the "Super" as pictured here, came out early in 1928 and was issued a type certificate number in July of that year. The "Super Universal" type was manufactured by the Atlantic Aircraft Corp. at Teterboro, N. J. which was a division of the Fokker Aircraft Corp. of America.

Listed below are specifications and performance data for the "Wasp" powered Fokker "Super Universal"; wing span 50'8", chord

Fig. 173. 1928 Fokker "Super Universal" with "Wasp" engine. The "Super" worked hard and led a romantic life in out-of-way places.

Fig. 174. A "Super Universal" on skis in the Canadian "brush-country".

is tapered in plan-form and section, wing area 370 sq. ft., airfoil "Fokker", length 36'7", height 8'11", empty wt. as landplane 3000, useful load 2150, payload 1250, gross wt. 5150 lb., max. speed 138, cruise 118, land 56, climb 950, ceiling 18,000 ft., gas cap. 120 gal., oil 12 gal., range 675 miles. Average price at factory in 1928 was $17,500. The following figures are for seaplane with "Hamilton" metal floats; empty wt. 3550, useful load 1600, payload 700, gross wt. 5150 lb., max. speed 130, cruise 110, land 55, climb 900, ceiling 16,000 ft., range 600 miles. Later in 1928 the price for landplane had risen to $19,540 and price as seaplane was $22,450. A later version of the "Super Universal" with a 450 h.p. "Wasp" engine was listed as follows; empty wt. 3250, useful load 2300, payload 1380, gross wt. 5550 lb., max. speed 138, cruise 118, land 59, climb 830, ceiling 17,500 ft., gas cap. 126 gal., oil 12 gal., range 650 miles. Price at the factory had gone up to $21,800. The fuselage framework was built up of welded chrome-moly steel tubing, lightly faired to shape and fabric covered. The one-piece full cantilever wing was built up of laminated spruce spars and spar flanges and plywood ribs, reinforced with plywood stringers and completely covered with plywood veneer. The early "Supers" had the rakish, pointed type canopy for the pilot's section as first used on the C-2 type Fokker "Tri-Motor", but the later models had a cleaner and simpler type as shown here in various views. The fabric covered tail-surfaces were also built up of welded chrome-moly steel tubing, and both movable surfaces had hinge over-hang or "aerodynamic balance"; the fin was ground adjustable and the horizontal stabilizer was adjustable in flight. All controls were cable operated and most of the cables were out in the open for ease of inspection and maintenance. Wheel brakes, metal propeller, and inertia-type engine starter were available. The "Super Universal" was built on this "type number" clear into it's last days, sometime in the early "thirties". For the next "Fokker" development, the "Tri-Motor" F-10, see chapter for A.T.C. #56 in this volume.

Fig. 175. 1929 Fokker "Super Universal" in service with American Airways, originally operated by Universal Air Lines.

A.T.C. #53
(7-28)
"COMMAND-AIRE" (OX-5), 3C3

Fig. 176. The handsome "Command-Aire" was well-behaved and offered good performance with the OX-5 engine.

The very first of the popular "Command-Aire" 3C3 series (X-3790) was introduced by the Arkansas Aircraft Co. at Little Rock, Ark. about Jan. in 1928; and though it's introduction to the industry created somewhat of a mild stir and it was later well accepted into flying circles about the country, it was not actually a very great departure from the similiar commercial airplanes of this type already in existance. However, in it's favor was the fact that it was a well thought out design of good aerodynamic arrangement. By the very nature of it's well planned configuration, it proved to be inherently stable from all attitudes and flew very well with a light and sharp control.

Rightfully proud of their airplane's ability, and overwhelmed by enthusiasm, the company loved to bally-hoo it's exceptional handling and responsive stability by having their fearless test-pilot, a brave soul named Wright "Ike" Vermilya, to actually ride a-straddle of the fuselage while the airplane was flying by itself! This, to graphically prove that their claims were justified and that their airplane actually could fly "hands off". Without a parachute this was foolhardy perhaps, but a very effective demonstration indeed. News about a new airplane type traveled like wildfire to all corners of the country in this stage of aviation development, and although already well known about the country by name; seeing an "OX-5 Command-Aire" was still a rare sight and a treat to many for some time.

Not in the least bit discouraged by the large number of airplanes of this particular type already being manufactured, the "Command-Aire" was designed and developed late in 1927 by Albert Voellmecke. Albert Voellmecke was formerly with the Ernst Heinkel Airplane Works of Germany, and his talent and concepts acquired abroad was clearly reflected in the "Command-Aire" design. In it's basic form, the 3C3 was a 3 place open cockpit biplane that was powered with an 8 cylinder Curtiss OX-5 engine of 90 h.p. It had pleasant flight characteristics and it's performance on "90 horse" was quite good; especially in the lower speed ranges where positive and near-absolute control of the airplane was possible, right down through the "stall". Enthusiastic pilot reports and comments were sufficient to confirm this time and again. The wings were of equal span with a generous amount of wing area, and the later production models had long span slotted-hinge ailerons on the lower wing panels; these ailerons were one of the features largely responsible for this airplane's effective low-speed control. The OX-5 engine was neatly and completely cowled in with removable metal panels, and the radiator was blended nicely into the nose-section. The robust landing gear was of the more or less typical split-axle type of a rather extreme tread. The tail-skid was of a formed steel tube with a "hardened shoe", and it was rubber shock-cord sprung. Later models of the "Command-Aire" discarded this type of "skid" for one of the "spring-leaf" type; the spring-leaf design was actually the most efficient and trouble-free tail skid that was ever used. Other interesting features of the "Command-Aire" design are listed below in the description of construction and assembly details.

The OX-5 powered "Command-Aire" model 3C3, of the later type as shown here, received it's certificate number in July of 1928 and they were being built by the newly formed firm of Command-Aire Inc. at Little Rock, Arkansas; in a plant formerly used by an automobile manufacturer. R. B. Snowden Jr. was the Pres., with Albert Voellmecke as the Chief Engr.; the noted pilot J. Carroll Cone was later engaged as chief of sales. The exact number of these "Command-Aires" that were built is difficult to determine, but chances are that they were not too many because new or unused OX-5 engines were finally running out of stocks at this time and getting somewhat scarce. To offset this, the company experimented with various "radial" engines such as the "Siemens", the "Walter", and the Warner "Scarab" engine; a small number of these were built in 1928-29. Though the OX-5 "Command-Aire" was rather sparse in number, they were extremely popular and there is a good possibility that a few of these are still in existence somewhere, and may someday be rebuilt to fly again! For the next approved development in the "Command-Aire" series, see chapter for A.T.C. # 118, this discusses the Warner "Scarab" powered model 3C3-A.

Listed below are specifications and performance data for the OX-5 powered "Command-Aire" model 3C3; span upper and lower 31'6", chord both 60", wing area 303 sq ft., airfoil Aeromarine 2A, length 24'6", height 8'4", empty wt. 1410, useful load 790, payload 350, gross wt. 2200 lb., max. speed 100, cruise 85, land 36, climb 510, ceiling 9,500 ft., gas cap. 40 gal., oil 4 gal., range 440 miles. Price at the factory had reached $3350 in 1929. The prototype model had weights as follows; empty 1275, useful load 800, payload 365, gross wt. 2075, with only a slight variation in performance figures. The fuselage framework was built up of welded chrome-moly steel tubing, faired to shape and fabric covered. Quite novel was the one-piece metal turtle-back that was easily removable for inspection and maintenance to components in the rear section of the fuselage. With the use of metal engine cowling, metal cockpit cowling, and the metal turtle-back; actually less than half of the fuselage was fabric covered. The wing framework was built up of solid spruce spars and spruce and plywood built-up ribs, also fabric covered. Ailerons were of a welded chrome-moly steel tube framework, also fabric covered. The prototype "Command-Aire" had 4 ailerons, but the later production model had only two ailerons of the slotted-hinge type in the lower wing panels. Actuation of all controls was by push-pull tubes and bellcranks, no wires, cable, nor pulleys were used. The fabric covered tail-group was built up of welded chrome-moly steel tubing, the fin was ground adjustable and the horizontal stabilizer was adjustable in flight. The extra wide center-section panel was supported by the usual "N" struts, with an extra strut running from the top wing to a fitting on the lower fuselage at the landing gear station; this method eliminated cross-wires in the center-section bay and afforded clear unobstructed vision to the front. The extra wide landing gear had a tread of 87". Both cockpits were quite roomy, comfortable, and well upholstered. Sales for the various models in the "Command-Aire" series were handled by the Curtiss Flying Service in 1929.

Fig. 177. Effective demonstration of the "Command-Aire's" ability to fly "hands off"!

A.T.C. #54
(7-28)
BOEING, MODEL 40-C

Fig. 178. The 4 passenger Boeing model 40-C of 1928. Designed for shuttle-type service up and down the Pacific coast.

Held to limited production because of undecisive plans for this version, the Boeing Model 40-C was more or less a design to fit the need and a prototype version and logical forerunner to the improved and very popular Model 40-B-4. An airplane that was later used by the Boeing System, and numerous other air-lines with such great success. Pictured here in very good likeness, we see the Model 40-C as a quite large two-bay biplane that was typical of the "40 series" design in most all respects; in fact, it was basically little more than an improved Model 40-A with an enlarged cabin section that seated two extra passengers. Being assured that passenger air-travel by this time was slowly but surely increasing, Boeing engineers were obliged to devise this extra seating offered in the 40-C, to handle this much welcomed increase in air travel.

As mentioned before, the cabin section was enlarged and now seated four passengers, the compartment just ahead of the pilot's station had been converted to seat the extra passengers, but the pilot was still seated in an open cockpit in the aft section of the fuselage. More than likely as an economy measure for the shuttle-type service proposed for this airplane, the powerplant for the Model 40-C was the smaller and much cheaper to operate 9 cylinder Pratt & Whitney "Wasp" engine of 410-450 h.p., and performance characteristics were more or less comparable to the earlier Model 40-A. The 40-A, which carried only two passengers but also carried a sizeable payload of mail and cargo.

The Model 40-C was primarily developed and built for Boeing's Pacific Coast service, this route was the old "P.A.T." system which had been absorbed into Boeing's expanding net-work; the route served many towns up and down the coast from San Diego to Seattle. The type certificate number for the Model 40-C was issued in July of 1928, five of these were built up to that time and only ten of these were built altogether; nine were built for the Boeing System and one was built for National Park Airways. A portion of these were then later modified and converted into the 40-B-4 type by the installation of the larger and more powerful 9 cylinder Pratt & Whitney "Hornet" engine of 500-525 h.p. The Model 40-C was manufactured by the Boeing Airplane Co. at Seattle, Washington.

Listed below are specifications and performance data for the "Wasp" powered Boeing

Model 40-C; span upper and lower 44'2", chord both 79", wing area 547 sq. ft., airfoil "Boeing", length 33'4", height 11'8", wheel tread 88", empty wt. 3522, useful load 2553, payload 1453, gross wt. 6075 lb., max. speed 125, cruise 105, land 54, climb 720, ceiling 14,500 ft., gas cap. 140 gal., oil 12 gal., range 700 miles. Price at factory was $23,500. The fuselage framework was built up of welded chrome-moly steel tubing and braced in the aft section with steel tie-rods, it was faired to shape and fabric covered; although more than half of the fuselage was covered with metal panels. The wing framework was built up of spruce spars and spruce and plywood built-up ribs, also fabric covered. The 40-C was bonded and shielded for radio, and the wings were wired for lights. For the next Boeing development, the model B-1E flying-boat, see chapter for A.T.C. #64 in this volume.

A.T.C. #55
(7-28)
STEARMAN - C3B

Fig. 179. The Stearman C3B was the choice of many sportsman-pilots because of it's exceptional performance.

As the annals of bygone days in aviation are written, the "Stearman" C3B without any doubt will be remembered as a very proud airplane; proud of it's versatile ability and also of it's heritage. Altogehter, it has had a very commendable existance and must surely be considered as one of our all-time greats in early aviation. Well planned, it's a fact that the basic design was so excellent from the out-set that it remained the basis for every "Stearman" airplane that was ever built. Very much like a thorobred, it had first a subtle suggestion and then an unmistakable visible evidence of true class and good breeding; blessed with in-born attributes that had held up admirably throughout all of the successive models that were produced.

By nature, the "Stearmans" were extremely rugged in character and their unfailing dependability, especially "when the chips were down", was a by-word long known among the the folks that fly. As time went on, it's use and proven popularity as a sort of "Pony Express" on many of the early short haul feeder-lines of our growing transcontinental air-mail system, was more than likely one of it's greatest claims to fame and fond remembrance; but it was also very popular with the so-called sportsman-pilot of this day who could afford and loved a good airplane with plenty of dash and spirit. To own a "J5 Stearman" and bask in it's reflected glory, was just about the pinnacle of many a pilot's dreams and hopes. Even as a workaday airplane the "C3B's" performance, without reservation, was surely among the very best; they were a complete charm to fly, with spirit and crisp determination, yet well-behaved and extremely sure-footed.

As pictured here in the various views, the Stearman C3B was an airplane of uncomplicated lines, yet with a gentle and classic beauty. In it's basic form it was a 3 place open cockpit biplane that was powered with a 9 cylinder Wright "Whirlwind" J5 engine of 220 h.p.; this was a true marriage and happened to be a thoroughly compatible combination whereby one tended to show off the relative merits of the other. The "Stearman" as a type, was first introduced early in 1927 at famous "Clover Field" in Santa Monica, Calif., the factory at that time was located in nearby Venice. Early models built at the Venice plant were powered with the Curtiss OX-5 and the "Hisso" engines; when Stearman moved to Wichita late in 1927, they still built a few airplanes powered with the OX-5 and "Hisso" engines but the "Whirlwind-Stearman"

Fig. 180. This playful "Stearman" is a J4 powered C2B used by Varney Air Lines, and served regularly into 1929.

soon became the standard model in production. Their introduction to the trade was received enthusiastically to say the least and by their very nature they soon managed to acquire an earned reputation for performance and utility that was only equaled, but not hardly ever surpassed. Varney Air Lines was one of the first operators to use Stearman equipment on their scheduled routes, and many more carriers soon followed suit. Besides daily chores on the "mail routes" or obeying the whims whatever of some play-boy pilot, many C3Bs were used by flying-schools to train those going for a transport license and for advanced training in aerobatics.

The design and development of the "Stearman" biplanes was naturally carried out as a labor of love by Lloyd Stearman, with the very able assistance of Mac Short, who was a devoted associate and also a very capable engineer. Lloyd Stearman will also be remembered as the designer of the early "New Swallow" of 1924, the airplane that practically rung the "death knell" on all of the "Jenny types" of that day and blazed the trail for the early "commercial airplane"! Later, he also designed the first of the famous "Travel Air" biplanes. If for no other reason, either of these accomplishments were proof positive and would surely vouch for Lloyd Stearman's vision and sound ability in the matter of airplane design.

The type certificate number for the model C3B was issued in July of 1928 and it was manufactured by the Stearman Aircraft Co. at Wichita, Kansas. According to record, about 40 airplanes of all types were built by Stearman up to this time. A J5 powered C3B was flown in the 1928 National Air Tour by David Levy, but due to miscellaneous misfortune

Fig. 181. A Stearman C3B fitted for blind-flying trainer by American Airways.

he had to be content with 17th place. For discussion on the improved "C3MB", see chapter for A.T.C. #137.

In the latter "thirties", after the "Stearman's" mail carrying days were just about over, they became extremely popular for "crop dusting"; doing so well in this exacting chore that they became more or less the standard equipment for this job throughout the land! Many of the more recent "Stearmans", of the W.W. 2 "trainer" type, have also become more or less standard equipment for this job at the present; but it shouldn't be surprising to know that a good number of the "old Stearmans" are still dusting and spraying to this day.

Listed below are specifications and performance data for the J5 powered Stearman model C3B; span upper 35', span lower 28', chord upper 66", chord lower 54", wing area 297 sq. ft., airfoil "Stearman", length 24', height 9', empty wt. 1625, useful load 1025, payload 445, gross wt. 2650 lb., max. speed 126, cruise 108, land 41, climb 1000, ceiling 18,000 ft., gas cap. 68 gal., oil 8 gal., range 620 miles. Price at the factory was $8970.

The fuselage framework was built up of welded chrome-moly steel tubing, faired to shape and fabric covered. The wing panels were built up of solid spruce spars and spruce and plywood built-up ribs, also fabric covered. Ailerons were on the upper wings only and were operated by push-pull tubes that came out of the cockpit and up into the center-section panel where they were connected to torque tubes and bellcranks for positive actuation. The landing gear was of the outrigger type and had a tread of 90", individual wheel brakes were standard equipment. The fabric covered tail-group was also built up of welded chrome-moly steel tubing, the fin was ground adjustable and the horizontal stabilizer was adjustable in flight. Earlier models of the C3B used a "prop spinner" and some had no head-rest for pilot's cockpit; later models had a head-rest but no "spinner" was used, the nose cowling had been rounded off. The C3B was also offered as a seaplane on "Edo" floats, for details of this version see chapter for Grp. 2 approval numbered 2-124. For the rare "Hisso" powered "Stearman", model C3C, see chapter for A.T.C. #62 in this volume.

Fig. 182. The 1928 Stearman C3B, powered with "Whirlwind" J5 engine. One of the most versatile airplanes of this period.

A.T.C. #56
(7-28)
FOKKER "TRI-MOTOR", F-10

Fig. 183. The Stately Fokker F-10 was a familiar sight on the west coast, Western Air Express had many of these in service.

The illustrious Fokker "Tri-Motor", model F-10, was primarily a development that was built to the specifications suggested by Western Air Express, and it was first put into service on their Los Angeles to San Francisco route on May 26 in 1928. Pictured here in one of the views is "X-5170", this was their first F-10 delivered; the first of an order for ten airplanes of this type and it was actually in trial operation before it's type certificate was formally issued. The improved F-10 type was in development late in 1927 and was first introduced at the turn of the year in 1928; it was a larger, more powerful, and more modern version of the earlier "Whirlwind" powered "Tri-Motor". Itself a versatile airplane that had built up quite an impressive world-wide reputation, a reputation acquired by it's many historic and record-setting flights. First use of the early "Tri-Motor" as a commercial air-liner of any success, is depicted here by the inaugural flight of Key West to Havana service by Pan American Airways in Oct. of 1927. Their first ship (NC-53) shown here, was an F-7A-3M; "Pan American" had 3 of these F-7A type in service on their Key West, Florida to Havana, Cuba route since 1927 and one of these "Tri-Motors" was kept in almost continuous service as late as 1932.

In it's basic form, the improved model F-10 was still very much typical in most all respects, it was a somewhat larger airplane and was purposely designed for the transport of passengers on a fast schedule and with a good degree of comfort. Impressive in size and majestic in character, it was a 14 place high wing cabin monoplane with the unmistakable "Fokker" configuration and the familiar all-wood full cantilever wing. For greater utility and a better performance reserve over the rough terrain encountered on the Los Angeles to Frisco route, it was powered with 3 P & W "Wasp" engines of 400 h.p. each; the added power in this combination gave a performance that was quite excellent for a ship of this size and type. It was well planned and well laid out, quite comfortable and commodious, and could be operated at a fair profit with near-capacity loads. The F-10 type proved it's mettle early, and was soon to become very popular on a number of other air-lines about the country; especially so in the later improved F-10A version, some of which were still in use regularly on scheduled flights as late as 1935. The type certificate number for the "Wasp" powered Fokker F-10 was issued in July of 1928 and it was manufactured by the Atlantic Aircraft Corp., a division of the Fokker Aircraft Corp. of America, with plants at Hasbrouck Hts. and

Fig. 184. This was the first Fokker "Tri-Motor" F-10 delivered to Western Air Express and put into service on May 26 in 1928.

Fig. 185. A Fokker "Tri-Motor" of the C-2 type, shown here on it's inaugural flight from Key West to Havana in Oct. of 1927. This short hop of 92 miles grew into one of the world's largest airline systems.

Teterboro, New Jersey.

"Fokker" Tri-Motor history was so interesting and colorful that it's almost imperative that we mention at least some of the hi-lights of achievement. The very first of the "Fokker" Tri-Motors, which was also the first tri-motored monoplane in the U.S., was introduced in this country in the year of 1925. Basically, it was a development from the standard single-engined model F-7 and was made into a "tri-motor" simply by powering it with 3 Wright "Whirlwind" J4 engines of 200 h.p. each; one engine was mounted in the nose and one engine was suspended under the wing in strut braced nacelles on each side of the fuselage. It had a rather stark interior but had ample seating for 10 people and delivered a performance that was quite good for a ship of this size. This first model was generally designated the F-7-3M, and was flown by the inimitable A.H.G. "Tony" Fokker and co-piloted by E. P. Lott in the first Ford Reliability Tour of 1925. The jovial Robert Noorduyn, Fokker's chief engineer, was a passenger and acted as part of the required payload. Needless to say, the airplane was a novelty and a big hit throughout the tour, especially due to "Tony" Fokker's remarkable shenanigans with the big ship to demonstrate it's agility and especially due to his many speeches given of it's praises! As a good many participants of that year's tour often remarked, it would have been better called "Fokker's Popularity Tour"!

The following year, Com. Richard E. Byrd used this "Tri-Motor" on his North Pole flight in May of 1926, thus the "Fokker" became the first airplane to fly over the northern pole. Capt. Geo. Hubert Wilkins used one of these F-7-3M and a "Liberty 12" powered F-7 also on his Arctic exploration flights in 1926. A year or so after his expedition of 1926, his "tri-motor" after some rebuilding from the remains of two airplanes, became the famous "Southern Cross" which was flown clear across the Pacific Ocean to Australia by Charles Kingsford-Smith in June of 1928. A U.S. Army "Fokker" Tri-Motor of the C-2 cargo-type, powered with 3 Wright "Whirlwind" J5 engines, was flown by Air Corps Lts. Maitland and Hegenberger on a flight from Oakland, Calif. to Hawaii. This flight of June in 1927 was the first venture across the Pacific from our shores. Com. Byrd used a similiar type ship on his trans-

Fig. 186. This first "Fokker" Tri-Motor of 1925, flew over the North Pole with Com. Byrd in May of 1926.

Atlantic flight from New York to Ver-sur-mer in France. His airplane was named the "America" and after hours of flying in impossible weather over the French mainland, it crash-landed in the surf just off the coast in June of 1927. Amelia Earhart, the celebrated woman-pilot, flew as part of the crew in the "Friendship", a J5 powered "Fokker" Tri-Motor on twin-floats which flew across the Atlantic Ocean to Ireland in June of 1928; thus she became the first woman to fly across the Atlantic Ocean successfully. Ruth Elder's earlier attempt with Geo. Haldeman in Oct. of 1927 wound up in the drink a good deal short of the tiny Azores Islands, luckily, they were picked up by a passing steamer! The U. S. Army's "Question Mark" was another record-setter, and so we could go on. Continental Motors was at the same time pioneering the "corporate airplane" with a "Fokker" Tri-Motor by shuttling it's executives between their Detroit and Muskegon plants. About 9 of these "Whirlwind" powered "Tri-Motors" were reported built in 1927 and possibly all but one or two of them were famous record-breakers. This then was the illustrious background history inherited by the "Fokker" F-10, a proud heritage and a wealth of experience that was bound to ensure a cinch chance for a successful start to any airplane type. The model F-10 of course, was destined to follow a much less glamorous though very useful career; a career that played an important part in the development of efficient air transportation.

Listed below are specifications and performance data for the "Wasp" powered "Fokker" Tri-Motor model F-10; wing span 79'2", chord tapered both in plan-form and section, wing area 854 sq. ft., airfoil "Fokker", length 49'11", height 12'5", empty wt. 7390, useful load 5110, payload 2450, gross wt. 12,500 lb., max. speed 135, cruise 110, land 65, climb 1000, ceiling 16,500 ft., gas cap. 360 gal., oil 30 gal., range approx. 700 miles. The fuselage framework was built up of welded chrome-moly steel tubing, faired to shape and fabric covered except for the forward portion. The continuous piece internally braced full cantilever wing panel was built up of laminated spruce spars, spar flanges and mahagony plywood ribs, reinforced with plywood stringers and covered completely with a plywood veneer skin. The fabric covered tail-group was also built up of welded chrome-moly steel tubing, the fin was ground adjustable and the horizontal stabilizer was adjustable in flight to correct for variations in loading. The robust landing gear was rubber "shock-cord" sprung and this was rather unusual for a ship of this size; later models had an "oleo" type landing gear. Three-bladed metal propellers, engine starters, and wheel brakes were standard equipment. After being in service for a length of time, many of these "F-Tens" were modified to conform to the latest F-10A type by the addition of tail-wheels to replace the original "tail-skid", they had oleo-type landing gears installed and some were equipped with low-drag "Townend ring" engine cowlings. The next development of the "Fokker" Tri-Motor was the "Super Tri-Motor", or model F-10A, for a discussion on this type see chapter for A.T.C. #96 in this volume.

Fig. 187. A modified version of the "Fokker" F-10 Super Tri-Motor, note change in pilot's cabin. This ship once owned by Shell Oil Co.

Fig. 188. Fokker Tri-Motor "Question Mark" which set endurance record by staying aloft for over 150 hrs. in 1929.

A.T.C. #57
(8-28)
ALEXANDER "J5 EAGLEROCK", A-1

Fig. 189. 1928 "Eaglerock" A-1 powered with "Whirlwind" J5 engine of 220 h.p. It's performance in mountainous country was exceptional.

This rakish looking airplane was the proud offering of Alexander Aircraft in their new "A series" that were introduced early in 1928. These soon became known as and generally referred to as the "New Eaglerock", or more often just as the "center-section Eaglerock". They were modified extensively and improved quite a bit but they still retained most of the familiar "old Eaglerock" characteristics such as a generous amount of wing area, with a considerable amount of interplane "gap" between the wings, and the general all-round robust and rugged appearance. The planform of the wing panels was now rounded off at the tips, and the strut arrangement of the wing truss incorporated a large "center-section" panel into the upper wing; this method was a marked improvement and made into a much better and stiffer truss than was on the earlier "Eaglerock". Especially better than those of the "long wing" variety which were known to flex under certain extreme conditions and because of this were jokingly called "ol' rubber wings". The center-section panel on the new "A series" had a span of 6 feet and on some models it held a fuel tank that in conjunction with the normal fuselage tank, provided a fuel supply of 70 gallons. All four wing panels were the same except for fittings, etc. The distinctive and familiar "long wing" feature of the early "Eaglerock", namely, the greatest span of the lower wing, was now discarded and the upper wing had the greatest span. Although the 1927 "combo-wing" had the greatest span in it's upper wing, the "old Eaglerock" was most often thought of as a "long wing" type. The cockpits on the new "A series" were also redesigned for roominess and better protection against the weather, and the fuselage was faired out much deeper to a better shape. A performance improvement in the new series was definitely noticeable.

In it's basic form, the model A-1 was a rather large 3 place open cockpit biplane and was powered with a 9 cylinder "Whirlwind J5" engine of 220 h.p. which gave it quite a lively and impressive performance. As shown here in the various views, the model A-1's apparent bulk seemingly belied it's actual dexterity in the air! Not generally considered as an acrobatic airplane, nevertheless, the "Whirlwind-Eaglerock" did very well in this respect and was capable of the most beautiful "hammer-head stalls" that you'd ever see. It was truly a fine airplane in every respect, but the "J5 powered" combination was still quite expensive for the average flyer's pocketbook and it tended to remain somewhat of a scarce type in numbers built. The type certi-

Fig. 190. This early 1928 "Eaglerock" was the prototype for the A-1 series.

Fig. 191. 1927 "Eaglerock" Long-Wing with Wright J5 engine. Flown by Cloyd Clevenger in 1927 Ford Air Tour.

ficate number for the "J5 Eaglerock", the model A-1, was issued in August of 1928 and it was in continuous production into the latter part of 1929. Alexander Aircraft was pushing these "new Eaglerocks" rather hard and a good amount of production was coming off the line; they could be bought with just about any engine combination from the 90 h.p. "OX-5" clear on through to the 260 h.p. "Menasco-Salmson". Because of this, the "A series" went into at least 15 versions.

The "J5 Eaglerock" (A-1) turned out to be very popular, and especially so in the western part of the U.S. where it's high altitude performance came in good stead. It was even known to serve in the desolate regions of the Yukon. A number of these "Whirlwind-Eaglerocks" were still flying in the late "forties" and early "fifties", yet acceptable for many useful chores. It was learned that just up until recently a number of these "center-section Eaglerocks" were still flying and earning their keep, and earning it very well too, by "crop dusting", "crop spraying", and many other useful chores that even include "fish planting". The "A series" Eaglerock were manufactured in a new enlarged plant by the Alexander Aircraft Co. at Colorado Springs, Colo., a division of Alexander Industries. This facility was one of the largest "commercial airplane" plants of this period and had a potential of building one airplane every hour!

Listed below are specifications and performance data for the "Whirlwind" J5 powered "Eaglerock" model A-1; span upper 36'8", span lower 32'8", chord both 60", wing area 336 sq. ft., airfoil Clark Y, length 23'11", height 9'10", empty wt. 1705, useful load 786, payload 340, gross wt. 2491 lb., max. speed 126, cruise 108, land 39, climb 1080, ceiling 17,900 ft., gas cap. (fuselage tank only) 46 gal., oil 7 gal., range 496 miles. A later version of the A-1 was typical except for the following figures; empty wt. 1705, useful load 986, payload 396, gross wt. 2691 lb., land 42, climb 1000, ceiling 17,000 ft., gas cap. (fuselage and center-section tank) 70 gal., range 650+ miles. Price at the factory averaged at $7500, or it could be bought for $2250 less engine and prop. In many cases the customer furnished his own engine and "prop", which were shipped to the factory for installation. On the "A series", the fuselage framework was built up of welded chrome-moly steel tubing, heavily faired to shape with wood fairing strips and fabric covered. Basically identical, the wing panels were built up of solid spruce spars that were routed to an "I beam" section, with spruce and plywood built-up ribs also fabric covered. The interplane struts were chrome-moly steel tubing of streamlined section, the interplane bracing was of streamlined steel wire. The fabric covered tail-group was also built up of welded chrome-moly steel tubing, the fin was ground adjustable and the horizontal stabilizer was adjustable in flight. An inertia-type engine starter, metal propeller, and wheel brakes were available. The wings were wired for lights. Alexander Aircraft had built 450 "Eaglerocks" of all types by July 1st of 1928, and this was only the beginning; many, many, more were yet to be built.

Fig. 192. The center-section "Eaglerock" was offered in any engine combination from 90 to 260 h.p. Ship shown here had 230-260 h.p. Menasco-Salmson engine.

A.T.C. #58
(8-28)
ALEXANDER "OX-5 EAGLEROCK", A-2

Fig. 193. 1928 "OX-5 Eaglerock" model A-2. Those who lost their hearts to the charms of the early "Eaglerock", loved this one even more.

The model A-2 shown here, was the bread and butter model in the "A series" Eaglerock line and was a companion model to the J5 powered A-1, and others. It was typical in most all respects except for the engine installation which in this case was the ever-popular but slowly dwindling away, Curtiss OX-5 engine of 90 h.p. Every one of the various and numerous models in the new "A series" were three place open cockpit biplanes and were basically typical from the firewall back; the engine installation and any modifications deemed necessary for a particular combination was the only difference, and this is what determined the model designation. As noted previously, any approved engine from 90 to 260 h.p. could be installed at the factory.

Typical of it's predecessor in the years back, the OX-5 powered model A-2 was a gentle and amiable airplane of quite good performance and pleasant flight characteristics; it was now a much better looking airplane and didn't seem so bulky and awkward. Those who lost their hearts to the charms of the early "Eaglerock", loved this one also, and even more. It's popularity ran high among the small operators and it was also used by numerous large flying-schools as a primary training ship. The only deterent to it's continuing popularity was the sorely depleted stock of new OX-5 engines, which had finally run out after all these many years! Planning to utilize the large number of "used OX-5" engines still in the hands of many, the "Eaglerock" was also sold less engine for $2250, and the customer could send his own engine to the factory for installation; this price included the motor mount, the radiator, and the engine cowling, but did not include the "prop". These "Eaglerocks" of 1928-29 were sold on time payment plan also and Alexander was the first American manufacturer to offer this convenience to the flying public as far back as 1926. The new Alexander Aircraft Co. plant was now at Colorado Springs, Colo. which was a ways out of Denver, and was one of the largest plants engaged in the building of commercial airplanes during this period. The type certificate number for the OX-5 powered "Eaglerock" A-2 was issued in August of 1928 and many hundreds were built. See chapters for A.T.C. #7 and #8 in this volume for a review of the early "Eaglerock" of the Combo-Wing and Long-Wing type.

Listed below are specifications and performance data for the OX-5 powered "Eaglerock" model A-2; span upper 36'8", span lower 32'8", chord both 60", wing area 336 sq. ft., airfoil Clark Y, length 24'11", height

Fig. 194. The "Eaglerock" A-2 had OX-5 engine, customers often shipped their own engines for installation at the plant.

9'8", empty wt. 1459, useful load 786, payload 340, gross wt. 2245 lb., max. speed 99, cruise 85, land 34, climb 514, ceiling 10,200 ft., gas cap. 46 gal., oil 4 gal., range 450 miles. Price at the factory with OX-5 engine, when available, was $2475, otherwise it was $2250 less engine and propeller. In Nov. of 1929 this price was lowered to $2000. The fuselage framework was built up of welded chrome-moly steel tubing, faired to shape with wood fairing strips and fabric covered. The cockpit cut-outs were not as pronounced and differed slightly from the A-1 and other models. The wing panels were built up of routed spruce spars and spruce and plywood built-up ribs, also fabric covered. The fabric covered tail-group was also built up of welded chrome-moly steel tubing, the fin was ground adjustable and the horizontal stabilizer was adjustable in flight. The vertical tail surfaces on the A series were more or less the same as on the early "Eaglerock", but the horizontal surfaces had been redesigned; all controls were cable operated. The standard color scheme for the A-2 was silver and blue, other colors were optional on special order. For the next development in the "A series", the "Eaglerock" A-3 and A-4, see chapter for A.T.C. #59 in this volume.

Fig. 195. This was the rare A-5, powered with Menasco-Salmson engine.

Fig. 196. The Alexander plant was one of the largest of this period, this view shows only a portion.

A.T.C. #59
(8-28)
ALEXANDER "HISSO-EAGLEROCK", A-3 & A-4

Fig. 197. This "Eaglerock" A-4 was flown by Cloyd Clevenger in 1928 National Air Tour.

The models A-3 and A-4 were also typical of the other "A series Eaglerock"; in basic form they were 3 place open cockpit biplanes and were powered with the war-surplus "Hisso" (Hispano-Suiza) engines, either the Model A of 150 h.p. or the hi-compression Model E of 180 h.p. As pictured here in very good likeness, the "Hisso-Eaglerock" made into a fine combination and was a very handsome airplane of good performance. Sharing the fate of most all the other "Hisso powered" airplanes, it's popularity was somewhat limited and it also lacked the general appeal to make it a big seller. Although, of all the airplanes that were built with the "Hisso" engines, there were probably more "Eaglerocks" than any other type. In all, some 93 "Hisso-Eaglerocks" were reported built.

The many refinements and changes built into the "New Eaglerock" series are clearly visible, the most striking changes were the rounded off wing tips, the robust looking wing struts, and the enormous center-section panel; all this provided better bracing and a more rigid wing truss, capable of great stress without deflection. The improved tail-group is clearly shown, along with a fuller fairing of the fuselage, and redesigned cockpits which offered more roominess and much better weather protection. There were four ailerons, one on each panel and they were torque tube operated for more positive control, the horizontal stabilizer was cockpit adjustable. One feature that definitely set the A-3 and A-4 apart from the "OX-5 Eaglerock" was the nose-type radiator; what the reason was for using a nose-type radiator on the A-3 and A-4 is doubtful, but it blended into the configuration very nicely and set it apart from the rest of the series. The OX-5 powered version had a free-air type radiator that was hung under the fuselage just ahead of the landing gear.

An A-4 type, powered with the 180 h.p. "Hisso E" engine, was flown in the 1928 National Air Tour by Cloyd Clevenger and finished in 18th place. Clevenger, by the way, was Alexander's chief pilot in charge of test and development for a number of years. An "A series" Eaglerock powered with a 9 cylinder Menasco-Salmson engine of 230-260 h.p. was the model A-5 and is also shown here. This ship was an added entrant in the National Air Tour for 1928 and finished in 25th place. As part of the "Eaglerock team" for that year, it was flown by the affable "Benny" Howard, an air-mail pilot later of air-racing fame. A fame brought about by his skill with his famous "DGA" airplanes; the

Fig. 198. 1928 "Hisso-Eaglerock" model A-4. The "Hisso" engine and the "Eaglerock" made into a compatible combination.

spectacular "Pete", "Mike" and "Ike", and of course "Mr. Mulligan"! The type certificate number covered both the models A-3 and A-4, and was issued in August of 1928. Manufactured by the Alexander Aircraft Co. at Colorado Springs, Colo. which was still enthusiastically guided by the paternal efforts of J. Don Alexander as Pres., with hard-working J. A. McInaney drummin' for sales, and Al Mooney handled the design and engineering.

Listed below are specifications and performance data for the "Hisso" powered "Eaglerock" models A-3 and A-4; span upper 36'8" span lower 32'8", chord both 60", wing area 336 sq. ft., airfoil Clark Y, length 24'10", height 9'10", empty wt. 1832, useful load 786, payload 340, gross wt. 2618 lb., max. speed 114, cruise 97, land 40, climb 720, ceiling 12,850 ft., gas cap. 46 gal., oil 5 gal., range 377 miles. These figures are for the A-3 with 150 h.p. "Hisso", with 180 h.p. "Hisso" they were identical except for the following figures; max. speed 120, cruise 103, land 40, climb 870, ceiling 14,000 ft., gas cap. 46 gal., oil 5 gal., range 330 miles. A later version of the A-3 and A-4 had a gas capacity of 70 gallons which increased the useful load figure and the gross wt. by some 145 lbs.; this caused a slight change in the performance figures and afforded a cruising range of 500-575 miles. Total gross weight was 2762 lbs. Price at the factory was $3250 with government-overhauled "Hisso A". Price in 1929 had gone up to $4000. For construction details of the A-3 and A-4, which were typical, see previous chapters. The next development in the "Eaglerock" A series was the A-7 that was powered with 125 h.p. "Siemens" engine, see chapter for Grp. 2 approval numbered 2-1 For the model A-12 with "Comet" engine, see chapter for A.T.C. #139.

Fig. 199. Popularity of the "Hisso-Eaglerock" rose rapidly when supply of OX-5 engines dwindled; the "Hisso" engines were still available in good number.

A.T.C. #60
(8-28)
SIKORSKY "AMPHIBION", S-38A

Fig. 200. 1928 Sikorsky "Amphibion" model S-38A, powered with 2 "Wasp" engines.

Even though this is the first certificated airplane to be built by Sikorsky, they certainly weren't newcomers in the airplane manufacturing business, by any means. Even in the U.S.A. the "Sikorsky" airplane dates back to late 1923 when the model "S-29", shown here in one of the views, was being built and then introduced at Roosevelt Field on Long Island. After preliminary checks and tests, it was successfully test-flown in Sept. of 1924 by Igor Sikorsky himself. A long time in the making, the S-29 was developed and built by Igor Sikorsky and his loyal little band of Russian refugees, who were capable artisans of all sorts that were patiently molded into builders of airplanes! Such an undertaking and the results obtained thereof speak very well for Igor Sikorsky; a gentle man of great intuition and engineering sense who has been one of the world's greatest inventors in the field of aviation. Their first airplane, the venerable S-29, roamed the country-side far and wide and made quite a name for itself as a cargo-carrier, and a cross-country charter transport. At one time during it's career, it was being piloted by the inimitable Roscoe Turner, long before he acquired any air-racing fame. Since that time forward, many different models were developed and built that preceded the model "S-38" featured in this chapter and all were very efficient and versatile airplanes, enjoying varying degrees of fame and success.

The model S-38A, shown here in various views, was an 11 place cabin sesqui-plane of the amphibious type, and it was powered with 2 Pratt & Whitney "Wasp" engines of 410 h.p. each. These can be seen mounted in strut braced streamlined nacelles placed between the two wings in a "tractor" fashion. The short and rigid boat-type hull which housed the cabin section was spacious and comfortable and quite seaworthy; for amphibious operation, a retractable wheeled landing gear was provided for ground landings. The outrigger booms which held the tail-group high out of the water and into the effective slipstream, were to become a distinctive "Sikorsky" trade-mark, and were somewhat reminiscent of the classic old "NC-4" flying-boat type. The Sikorsky "Amphibion" was indeed a majestic and unforgettable sight while in the air, and quite an impressive vehicle while on the ground, or in the water too.

The model S-38A was primarily developed for intended use by Pan American Airways and the N.Y.R.B.A. Line on their net-works that were shaping up in the Carribean Sea, and later into So. America. The S-38 type was

Fig. 201. The "Liberty" powered S-29, first "Sikorsky" airplane built in U. S. (1924).

a development from the earlier S-36 type. Also shown here in one of the views, the model S-36, which was Sikorsky's first amphibian type, was also developed for use by Pan Am because of the varied landing conditions that were to be encountered on their new net-work of routes among the Carribean Islands. One of the S-36 amphibian type was delivered to "Pan Am" in Dec. of 1927 and was in use until Feb. of 1928; in the meantime they were awaiting delivery of one of the S-38A type. The S-38A was an improved type that allowed a greater payload at a higher cruising speed and had a bit more cruising range. Records available disclose that 11 of the S-38A type were built; the N.Y.R.B.A. Line had the first one which was delivered in June of 1928, "Pan American" had 3 and one of these blazed a trail to Panama with Chas. Lindbergh; "Jock" Whitney used one as a luxurious air-yacht; the Curtiss Flying Service had 3 for charter-work and what-have-you; the U.S. Navy had 2; and Western Air Express had one that they used on their flights to and from Avalon, on Catalina Island. It became proven fact before long that these Sikorsky "Amphibions", in all types, were rugged and reliable in difficult service, and proved very functional despite their unusual layout. The greatest tribute to this airplane is voiced from the men who flew it!

The type certificate number for the model S-38A was issued in August of 1928 and they were manufactured by the Sikorsky Manufacturing Corp. at College Point, Long Island, New York. See chapter for Grp. 2 approval numbered 2-36, for S-38AH modification which was typical of the "A series" but was powered with 2 Pratt & Whitney "Hornet" engines of 525 h.p. each. The next development in the Sikorsky "Amphibion" line and also used extensively by Pan American Airways and others, was the improved model S-38B. See chapter for A.T.C. #126.

Back in the early "twenties", airplane manufacturers found it increasingly difficult to keep any semblance of development going, and to devise means of keeping that wolf away from the door; so like many others during this lean period, Sikorsky also built efficient, high performance wing panels on order, as replacements of the originals for

Fig. 202. The J5 powered S-36, America's first practical amphibian transport and the basis for the development of S-38 type.

Fig. 203. View showing pilots' cockpit and engine nacelles of S-38A.

types like the old mail-carrying "DeHaviland 4", the Curtiss "Jenny", the "Standard", the Curtiss "Oriole", and others. A classic example of this type of modification is the "Sikorsky" UN-4, which was basically a "Jenny" fuselage with a redesigned nose-section and with a parasol-mounted "Sikorsky" hi-lift wing. For further details, see chapter for Grp. 2 approval numbered 2-23 which will follow later. To give a summation and a brief description of the "Sikorsky" types that preceded the model S-38A, the listing is as follows.

As mentioned previously, the first of the "Sikorsky" type built in the U.S.A., was the S-29; this means that 28 types had already been built previous to this! The S-29 was a 16 place transport biplane that was powered with 2 "Liberty 12" engines of 400 h.p. each, it was the enclosed cabin type but the pilot and mechanic were seated in an open cockpit far aft in the fuselage. The model S-30 was a 10 place transport type cabin biplane with the pilot finally being placed under cover, and it was powered with 2 Wright "Whirlwind" J4 engines of 220 h.p. each. The model S-31 was a 3-4 place open cockpit biplane of the utility type and it was powered with a J4 engine; incidently, this was Sikorsky's first order for an airplane in the U.S., the order was placed by "Fairchild" and the ship was to be used in their aerial-surveys as a photo-ship. Fairchild had not yet become a manufacturer of airplanes! The model S-32 was a large 5 place open cockpit biplane that was powered with a 12 cyl. "Liberty" engine; this type was used for exploration work and operated with either wheels or floats, it saw extensive service in So. America. The model S-33, a very rare type, was called the "messenger" and was the smallest of the "Sikorsky" types; it was a 2 place open cockpit biplane and was powered with a 3 cyl. "Lawrence" engine of 60 h.p. This engine was the grand-daddy of the famous "Whirlwind" series. The model S-34 was a twin-engined experimental flying-boat type which provided the basis for the development of the S-36 amphibian. The model S-35 was the plane built for Rene Fonck's attempt to fly the Atlantic Ocean to Paris, it was a large cabin type transport biplane and was first powered with two engines but Fonck demanded modification to 3 engines, the plane crashed on take-off

Fig. 204. Interior of S-38A.

and the attempt was abandoned. The model S-36 was the first successful and practical commercial transport-type amphibian in this country and was primarily a development for use on such net-works as contemplated by "Pan American" and others; it was a flying-boat type amphibian of sesqui-plane arrangement and it had an enclosed cabin section in the hull which seated 8 places, it was powered with two "Whirlwind J5" engines of 220 h.p. each. The model S-38 was an improved development of this earlier type. The next in line was the model S-37 which was actually a modification of the former S-35 type that was built for Rene Fonck; after the crash, the remains of the plane were modified into a passenger transport that was powered with two Bristol "Jupiter" engines (British) of 500 h.p. each. Another of this type was a bomber-version that was powered with two "Hornet" engines of 500 h.p. each, this was the "Guardian" as tested by the Army Air Corps. This array of beautiful and versatile airplanes brings us on up to the model S-38A, as described somewhat briefly in this chapter.

Listed below are specifications and performance data for the "Wasp" powered Sikorsky "Amphibion" model S-38A; span upper 71'8", span lower 36'1", chord upper 100", chord lower 59", wing area 720 sq. ft., airfoil "G.S.1", length overall 40'3", height on wheels 13'10", wheel tread 116", height in water 10'2", empty wt. 6000, useful load 4480, payload 2800, gross wt. 10,480 lb., max. speed 120, cruise 103, land 55, climb 750, ceiling 16,000 ft., gas cap. 180-270 gal., oil 24 gal., range 500-750 miles. The hull structure was built up of oak and ash frame members that were reinforced with dural plates and gussets, outer covering was of heavy gauge dural sheet. The hull was arranged with numerous water-tight compartments for greater safety. Entry into the cabin was by way of a hatch in the rear section. The wing framework was of all metal construction and was fabric covered; all fuel tanks were mounted in the upper wing, allowing for a maximum fuel load of 300 gallons, oil tanks were also mounted in the upper wing for a capacity of 24 gallons. Landing gear retraction required about 50 seconds; to operate as a flying-boat, landing gear was removable from 6 mounting bolts. Each wheel was retractable separately and this was often a great help in taxiing in rough water and high winds. The S-38A had a sufficient performance reserve and could maintain level flight at 90 m.p.h. on only one engine. The next development in the Sikorsky "Amphibion" type was the S-38 B, see chapter for A.T.C. #126.

A.T.C. #61
(8-28)
FAIRCHILD, FC-2W2

Fig. 205. The Fairchild FC-2W2 was the answer to demands for increased payload and cargo-carrying capacity.

The continued success of the "Fairchild" monoplanes, especially in the "bush country" of the U.S. and Canada, was a welcomed justification for the type, but it also brought on greater demands for increased payload and more cargo carrying capacity. To meet and cope with this demand, Fairchild developed and brought out the improved model FC-2W2. It was a somewhat larger airplane than either the standard FC-2 or the FC-2W models, but it was still basically typical to both in most all respects. In it's basic form, the model FC-2W2 was a 5 to 7 place high wing cabin monoplane with the familiar strut braced semi-cantilever wing that could also be folded back as on the other earlier "Fairchild" monoplanes, but it now had a cabin space of much larger capacity; over 145 cu. ft., to handle bulkier loads and it could carry up to 7 passengers or more than 1500 lbs. of payload. The powerplant for this model was the 9 cylinder "Wasp" engine of 410-450 h.p.; it was capable of very good short-field performance on and over all sorts of terrain. Proving itself extremely suitable and quite popular as a working airplane, devoid of frills and bad temper; working under conditions that were usually anything but ideal.

The Fairchild FC-2W2 has an enviable and interesting history in the frontiers of early aviation where the going was always rough, whether it be in the U.S.A., Mexico, Canada, Alaska, So. America, or even in China and the "South Pole"! One of the accompanying illustrations shows an FC-2W2 on Fairchild-built "floats", flying high over the desolate waste lands of northern Canada. Com. Byrd's FC-2W2, named the "Stars and Stripes", and no doubt the most famous and best known of this type, was the first airplane to explore the "South Pole" regions in the Antarctic, in the year of 1928. After a good deal of exploration and mapping of this territory, it was put into "deep freeze", and about 5 years later on Byrd's return trip to the "pole", it was dug up and thawed out and put to use again! A Pan American Airways division used the FC-2W2 with great success on flights across the treacherous Andes mountains on their run from Buenos Aires, Argentina to Santiago, Chile. One of the views shown here pictures the typical terrain encountered on these flights; certainly a good test and recommendation for an airplane's ability.

So, pictured here in the various views, we see the model FC-2W2 in it's natural habitat, with all of it's familiar Fairchild traits and characteristics. Introduced earlier in the year, the type certificate number for this model was issued in August of 1928 and after a production of only a small number, it became the basis and the direct forerunner to the "Fairchild 71". For a detailed account of the "Model 71", see chapter for A.T.C. #89 in

Fig. 206. This FC-2W2 was a long time in Canadian service with Canadian-Colonial Air Lines. Note tail-wheel and Goodyear "air wheels" which were not used in 1928, but were added later.

Fig. 207. FC-2W2 on Fairchild metal floats, high over the desolate waste-lands of northern Canada.

this volume.

Listed below are specifications and performance data for the "Wasp" powered "Fairchild" model FC-2W2; wing span 50', chord 84", wing area 310 sq. ft., airfoil Gottingen Mod., length 33'2", height 9'6", wts. as landplane, empty wt. 2732, useful load 2768, payload 1526, gross wt. 5500 lb., max. speed 134, cruise 108, land 55, climb 875, ceiling 15,000 ft., gas cap. 148 gal., oil 12 gal., range 750 miles. Wts. as float-seaplane; empty 3072, useful load 2428, payload 1186, gross wt. 5500 lb., max. speed 127, cruise 104, land 55, climb 850, ceiling 14,500 ft., range approx. 700 miles. As a seaplane, the FC-2W2 was equipped with Fairchild metal pontoons. The fuselage framework was built up of welded chrome-moly steel tubing, lightly faired to shape with wood fairing strips and fabric covered. The wing framework was built up of spruce box-type spars and spruce and plywood truss-type ribs, also fabric covered. The fabric covered tail-group was built up of welded chrome-moly steel tubing, the fin was ground adjustable and the horizontal stabilizer was adjustable in flight. The seats were quickly and easily removable for clear floor space when carrying cargo, or sometimes a few seats were left in if an occasional passenger or two were making the same flight. It would be factual to say that the hard-working "Fairchilds" had served in the lands of the rugged individual, and it was not uncommon for a paying passenger to be sitting amongst a load of mining machinery or supplies. Just up until a few years ago there was an FC-2W2 still flying in and around the Pacific northwest, others are still around but probably not active.

Fig. 208. This FC-2W2 was the famous "Stars & Stripes".

A.T.C. #62
(8-28)
HISSO-STEARMAN, C3C

Fig. 209. 1928 Stearman C3C with "Hisso" engine, this was a rare combination.

Just about every manufacturer in the business have built an occasional "rare bird" or two, and even "Stearman" got around to building a few. The model C3C, shown here, was one of them; there were a few built it seems, but actually it was only considered as a rare type because the "Stearman" biplane was so thoroughly identified with the Wright "Whirlwind engine that it seemed out-of-place to see it with any other installation. The model C3C was basically typical to the model C3B in just about every respect, except that it was powered with the 8 cylinder water-cooled vee-type "Hisso" (Hispano-Suiza) engines. Available with either the Model A of 150 h.p. or the Model E of 180 h.p.

Besides hearsay and hangar-talk, little is known or recorded about this version, except that it's performance was very good, which is only typical of all "Stearman" aircraft. The type certificate number for the "Hisso" powered model C3C was issued in August of 1928; for a mention and brief discussion on other "Stearman" rare-birds, see chapters for Grp. 2 approvals numbered 2-53 for the model C2K, number 2-58 for the model C3L, and number 2-70 for the model C3K. The "rarest bird" of all the "Stearman" line at this point was no doubt the OX-5 powered version, also shown here, that was built in Venice, Calif. and to some extent in Wichita. The "Stearman" biplanes were now manufactured by the Stearman Aircraft Co. at Wichita, Kansas.

Listed below are specifications and performance data for the Hispano-Suiza powered "Stearman" biplane model C3C; span upper 35', span lower 28', chord upper 66", chord lower 54", wing area 297 sq. ft., airfoil "Stearman", length 24'6", height 9', empty wt. 1790, useful load 960, payload 380, gross wt. 2750 lb., (with Hisso E of 180 h.p.; max. speed 121, cruise 106, land 45, climb 900, ceiling 16,000 ft.) (with Hisso A of 150 h.p.; max. speed 116, cruise 100, land 45, climb 820, ceiling 15,000 ft.), gas cap. 68 gal., oil 5 gal., approx. range 550 miles. These figures have not all been confirmed but should come well within the normal tolerances. The fuselage framework was built up of welded chrome-moly steel tubing, faired to shape with wood fairing strips and fabric covered. The wing panels were built up of solid spruce spars and spruce and plywood built-up ribs, also fabric covered. The maximum fuel load of 68 gal. was carried in two tanks, one in the fuselage ahead of front cockpit, and one in the center-section panel of the upper wing. The fabric covered tail-group was built up of welded chrome-moly steel tubing, the fin was ground adjustable to offset high r.p.m. torque, and the horizontal stabilizer was adjustable in flight. The landing gear was of the typical

"Stearman" out-rigger type and had a tread of 90". Metal propeller and wheel brakes were available, wings were wired for lights. The next developments in the "Stearman" biplane line were the M-2 "Speed Mail" and the C3MB. For discussion on these, see chapters for A.T.C. #127 and #137.

Fig. 210. This "Hisso-Stearman" operated in the Pacific north-west.

Fig. 211. 1927 Stearman with OX-5 engine, this is one of the earliest examples of this type, and was built in Venice, Calif. plant.

A.T.C. #63
(8-28)
CURTISS, "CHALLENGER-ROBIN"

Fig. 212. The 1928 Curtiss "Challenger-Robin"

This "Robin" shown here, was the next development of the Curtiss series, this was the "Challenger-Robin"; an airplane that was typical to the earlier "OX-5 Robin" in most all respects except for the engine installation which in this case was the new 6 cylinder Curtiss "Challenger" engine of 165 h.p. The "Challenger" engine was an air-cooled radial type of a rather odd configuration, it was a staggered twin-row "radial" that was actually two banks of 3 cylinders each operating off of a two-throw crankshaft. Very similiar in principle to the old 6 cylinder "Anzani" engines. This arrangement produced an odd sound that was very distinctive; it gave off a rather rough sounding exhaust tone, but the engine was really quite smooth running. Performance records later set by this engine proved it's dependability and stamina beyond question, and it became very popular in this power range. The "Challenger" engine was designed by Arthur Nutt who was the chief engineer of Curtiss' motor division, and it went into a number of modifications and improvements before it was finally discontinued sometime in 1931 at 185 h.p.

To get back to the airplane in discussion here, the "Challenger-Robin" was also a 3 place high wing cabin monoplane of identical configuration to the earlier "OX-5 Robin" (see chapter for A.T.C. #40 in this volume), but the added power of the 165 "horse" Challenger engine produced a considerable increase in all-round performance and utility. This model was also a development by the Curtiss parent company and was introduced just a short time after the "OX-5 Robin". After sufficient testing and some modification, it was also turned over to the Curtiss-Robertson subsidiary for production. See chapter for A.T.C. #69 in this volume for discussion on the production version. The type certificate number for the Curtiss "Challenger-Robin" was issued in August of 1928.

This "Robin" shown here, also had the characteristic "airfoiled struts" that contributed 41 sq. ft. to the lifting area, and helped to increase the stability of the wing. The exhaust collector-ring was draped around the outside of the engine cowling, on later versions it was buried on the inside; the cowling also lacked the numerous louvers that were used on later models for venting the engine compartment. The cabin seating arrangement was retained as on the earlier "Robin", but the third door that was in the rear L.H. side was eliminated. First tested with rubber discs in the landing gear shock-absorbing system, this was later discarded in favor of the "spring and oleo" type. The tail-skid also swiveled and was sprung with Rusco rubber discs, there were no wheel brakes. There were two fuel tanks of 25 gallons each, one in each wing root for a total supply of 50 gallons.

Listed below are specifications and performance data for the "Challenger" powered Curtiss "Robin"; wing span 41', chord 72", wing area 224 sq. ft., airfoil Curtiss C-72, length 24'1", height 7'10", empty wt. 1576, useful load 864, payload 394, gross wt. 2440 lb., max. speed 118, cruise 102, land 47, climb 640, ceiling 12,500 ft., gas cap. 50 gal., oil 5 gal., range 500 miles. The fuselage framework was built up of welded chrome-moly steel tubing, lightly faired to shape and fabric covered. The wing framework was built up of routed spruce spars and Alclad aluminum stamped-out ribs, also fabric covered.

Fig. 213. The "Challenger-Robin" was a Curtiss development that was later built by Curtiss-Robertson subsidiary.

A.T.C. #64
(8-28)
BOEING "FLYING BOAT", B-1E

Fig. 214. Boeing model B-1E powered with "Wasp" engine.

There was a time when the "flying boat" type of airplane was very popular and even quite numerous, because water abounds almost everywhere and it provides ready-made landing fields even right in the "heart of town", so to speak; therefore offering a convenience and utility that could not be equalled in a land-plane type of airplane. This advantage however became more or less nullified with the advent of the "amphibian" and "float-seaplane" types, making the flying-boat somewhat less desirable and relegated to serve only on specialized jobs, or in certain areas of the country.

The Boeing model B-1E was basically an improved and more powerful version of the earlier model B-1D, which proved itself so useful and had been put to work in the Pacific northwest and in parts of Canada. In it's basic form, the model B-1E was a cabin biplane of the "flying boat" type; it had roomy seating for four and it was now powered with a 9 cylinder Pratt & Whitney "Wasp" engine 410-425 h.p. The "Wasp" was mounted in a streamlined nacelle that was hung between the wing panels in a "pusher" fashion; needless

Fig. 215. This was the "Liberty" powered B-1 of 1919. In over 9 years of service it wore out seven engines and traveled more than 500,000 miles.

to say, the added power of the "Wasp" did wonders for this ship and for an airplane of the flying boat type, it had a very lively performance.

Though very limited in number, the Boeing "flying boats" led a very active life and performed many a diversified service. For instance; they transported miners and supplies from seaboard areas to lakes in the interior, participated in the "dusting" of pest-infested forests, were used for "spotting" schools of fish and the hauling of cannery supplies. They were supplying taxi-service from Alaska to various points in Canada clear down to British Columbia; were prospecting in new mining areas, and even carried the air-mail. One Boeing "boat" of a later model was engaged in giving "joy-riders" a sight-seeing tour around Catalina Island, just off the coast of California.

Going farther back in the Boeing "flying boat" series, there was a "Liberty" powered B-1, shown here, that was in active service for over 9 years, from 1919 to 1928 and wore out seven engines! In this time, it traveled more than 500,000 miles on an international air-mail run between Seattle, Wash. and Victoria, B. C., truly an outstanding record of service and durability. Of this kind of stuff the B-1E was made! The type certificate number for the model B-1E was issued in August of 1928. The B-1E was later modified into a 6 place version and this became the basis for the "Model 204". For discussion on the Model 204, see chapter for A.T.C. #157.

Listed below are specifications and performance data for the "Wasp" powered Boeing flying-boat model B-1E; span upper and lower 39'8", chord both 79", wing area 466 sq. ft., airfoil "Boeing", length 32'7", height 12'2", empty wt. 2990, useful load 1510, payload 800, gross wt. 4500 lb., max. speed 125, cruise 103, land 57, climb 1200, ceiling 13,500 ft., gas cap. 80 gal., oil 10 gal., range 500+ miles. Price at the factory was approx. $20,000. The hull framework was built up of spruce members that were covered with mahagony plywood. The wing framework was built up of solid spruce spars and spruce and plywood truss-type ribs, also fabric covered. The fuel supply was carried in the upper wing. All movable control surfaces were cable operated, the horizontal stabilizer was adjustable in flight. Engine starter and metal propeller were available, the wings were wired for lights.

Fig. 216. Eddie Hubbard (left) and Wm. E. Boeing in front of a Boeing C-700 on a survey flight of Mar. 3, 1919. This was first flight on international mail route from Seattle to Victoria, B. C. in Canada.

Fig. 217. Boeing flying-boat B-1E, a versatile type that performed many varied services. Tho' gone from the scene, the romance of the "flying boat" will be well remembered.

A.T.C. #65
(8-28)
CESSNA, MODEL A (AA)

Fig. 218. The 1928 Cessna "Model AA", powered with 10 cyl. Anzani engine of 120 h.p. This design was so excellent that it remained the basis for Cessna aircraft development in the next 25 years!

This slender and vivacious looking little lady, that is pictured here, was an airplane that created quite a stir among the flying folk in flying circles about the mid-west, when it was first introduced as a prototype in the latter part of 1927. It was one of the very first light commercial airplanes to make practical use of the internally braced wing, and the first to offer a high performance without the sacrifice of comfort or utility, on a minimum of power. Nearly a year later, in a slightly modified version, it became the first certificated model of the very popular and certainly quite famous "Cessna Cantilever Monoplane" series. In it's basic form, it was a trim 3-4 place fully enclosed high wing monoplane using a one-piece full cantilever wing; a configuration calculated to keep parasitic resistance, rigging, and maintenance, down to a bare minimum. Mainly because of the lack of suitable American engines in this power range, or perhaps because of Cessna's familiarity with the type; it was powered with a 10 cylinder "Anzani" engine. The "Anzani" (French) was a 10 cylinder air-cooled "two row radial" engine of 120 h.p., an engine that was quite easily available in this country but not always too popular, more than likely due to it's messy and somewhat cranky habits. Still...Clyde V. Cessna, fondly known as the "old master", managed to get a performance far above average, even with this combination. The high performance, utility, and efficiency, of Cessna's "cantilever monoplanes" has been well established, and has been well known to a good many throughout the years; it is not even surprising that this basic configuration with only slight changes and some occasional refinement, had stood the test of service and public taste for some twenty-five years!

The Cessna "Model AA" powered with the 120 h.p. Anzani engine, as pictured here; received it's approval for a type certificate number in August of 1928 and according to factory records, some 117 airplanes of this type were built. Besides this "Anzani" powered version of the basic Model A, it was also offered with the 9 cylinder Siemens-Halske engine (German) of 125 h.p. as the model AS, with the 7 cylinder "Comet" engine of 130 h.p. as the AC, with the 7 cylinder "Floco" (Axelson) of 115-150 h.p. as the AF, and with the 9 cylinder Wright J5 engine as the BW. For a descriptive discussion on these various versions of the basic "A series", see chapters for Grp. 2 approvals numbered 2-7 and 2-8 which will follow.

These models were more or less an experiment to find a suitable replacement for

Fig. 219. Prototype for Cessna "A" series, built by Cessna-Roos Aircraft Co. in 1927.

the "Anzani" engine, but these models failed to do this and were short-lived. All were soon replaced in production schedules by the "Cessna AW" which became the standard production model through most of 1929. The Model AW was basically typical in most all respects except that it was powered with the spunky little 7 cyl. Warner "Scarab" engine of 110 h.p. This version allowed for the seating of four people in neat but rather chummy quarters, with a performance on 110 horsepower that has never been equalled. For a discussion of this model, see chapter for A.T.C. #72 in this volume. The real "wildcat" of this series was the model "BW", which was a sportsman's version of the "A" that was powered with the 220 h.p. Wright J5 engine. It was capable of a terrific performance that could best be described as absolutely sensational.

Not particularly acceptable as is, the "Anzani" engines installed in the Model AA version, were modified and throughly reconditioned by Cessna to the point where they were quite reliable and had none of the objectionable characteristics of the original French "Anzani", which were best described by the words of one as "that awful contraption"! The "Cessna Cantilever Monoplanes" were manufactured by the Cessna Aircraft Co. at Wichita, Kansas with the inimitable Clyde V. Cessna as it's president and chief engineer.

Clyde V. Cessna's aeronautical activities date back to 1911, when he built his first airplane and also taught himself how to fly it. This first model was more or less a copy of the French "Bleriot" monoplane, but his second airplane, which was built in 1912 and is pictured here, was quite an improvement and offered much better performance and flight characteristics. He flew it quite frequently at aerial exhibitions. Always a staunch advocate for the monoplane type, Cessna had already built some 15 airplanes of his own design prior to 1925 when he joined forces with Walter Beech and Lloyd Stearman to help form "Travel Air". As time went on, there was evidence of frequent bickering between Cessna and Beech on the relative merits of the monoplane versus the biplane! This eventually caused a break between them, and Clyde Cessna elected to go out on his own again. Victor Roos, joined forces with Cessna in August of 1927 to form the Cessna-Roos Aircraft Co. at Wichita, Kansas. Shortly after, Roos pulled out for other interests, and Clyde Cessna formed the Cessna Aircraft Co. at Wichita, Kansas in Dec. of 1927. Their first models were known as the "A series" and were built on through 1929.

Listed below are specifications and performance data for the "Anzani" powered "Cessna" model AA; wing span 40', chord at root 86", chord at tip 58", (M.A.C. 66'), wing area 224 sq. ft., airfoil "Cessna" (Mod. M-12), length 24'9", height 7'2", empty wt. 1250, useful load 720, payload 340, gross wt. (as 3 place) 1970 lb., max. speed 120, cruise

102, land 40, climb 720, ceiling 10,000 ft., gas cap. 35 gal., oil 4.5 gal., range 450 miles. Price at the factory was $5750 with "Hamilton" wood propeller, later the price was raised to $6300 with wood prop and $6500 with metal prop. The "Anzani" engine as modified and completely rebuilt by Cessna, cost $2167 less propeller. The following figures are for the latest version of the model "AA", all measurements remained the same, only changes were; empty wt. 1304, useful load 956, payload 510, gross wt. (as 4 place) 2260 lb., max. speed 120, cruise 102, land 45, climb 670, ceiling 9,500 ft., gas cap. 40 gal., oil 4.5 gal., range 500+ miles.

The fuselage framework was built up of welded chrome-moly steel tubing, lightly faired to shape and fabric covered. The wing framework was built up of laminated spruce spars and spruce and mahagony plywood built-up ribs, also fabric covered. The full cantilever wing was built up in one continuous piece and was tapered in plan-form and section, it was bolted directly on top of the fuselage. The forward occupants used the front spar as a head-rest. The landing gear was built up of welded chrome-moly steel tubing streamlined in section, and was of a novel split-axle type using two spools of rubber shockcord to snub the taxiing and landing loads. The tail-skid was also rubber shock-cord sprung. Wheel brakes were available on later models. The adjustable stabilizer was fastened to rotate at the front spar and negative or positive adjustments were made by a novel screw-type jack permitting compensations for load variations while in flight. There were two fuel tanks of 20 gallons each, on either side of the fuselage in the wing's center section.

Fig. 220. Clyde V. Cessna and his second airplane, built by him in 1912.

A.T.C. #66
(8-28)
LOENING "AIR YACHT" (HORNET)

Fig. 221. 1928 Loening "Air Yacht" with 500 h.p. "Hornet" engine. Demands for increased performance led to the development of this version.

The next development of the commercial cabin-type Loening "Amphibian" series, was one more or less of necessity and was basically typical to the previous design discussed here under A.T.C. #34 in this volume. Excepting, that this new model was now powered with the larger and more powerful Pratt & Whitney "Hornet" engine of 500 h.p., and the fuselage was enlarged and redesigned to a better lay-out at the cabin section. The added power of the "Hornet" giving it a much better all-round performance and utility; it was also of 5 to 7 place capacity, but was now able to handle a load of 6 passengers and a pilot with much more verve.

As pictured here in the upper view, it was a rather large biplane of the two-bay wing panel type; 4 to 6 passengers were seated behind and below in the depths of the fuselage in an enclosed cabin section and the pilot was seated in an open cockpit up forward. The trussing of the two-bay wing panels was found necessary, no doubt, due to the comparatively thin wing section and the very large span. The "Hornet" engine swung a 3-bladed metal propeller and the exhaust collector-ring had a long tail-pipe which dispelled gases at a safe level over the top wing.

This passenger carrying type of the Loening "Amphibian" was first formally shown at the Detroit Aircraft Show in April of 1928. Demonstrations, tests, and consultations with prospective buyers, led to the design of this more powerful version and it was not too long before the Loening "cabin amphibian" was seen in service on various air-routes, and as private "air yachts". The type certificate number for this model was issued in August of 1928 and it was manufactured by the Loening Aeronautical Engineering Company in New York City.

Listed below are specifications and performance data for the "Hornet" powered Loening "Cabin Amphibian"; span upper and lower 46'8", chord both 72", wing area 517 sq. ft., airfoil "Loening", length 34'8", height wheels down 13'2", height wheels up 11'5", empty wt. 3867, useful load 2033, payload 1000, gross wt. 5900 lb., max. speed 112, cruise 100, land 52, climb 850, ceiling 13,500 ft., gas cap. 140 gal., oil 12.5 gal., range approx. 500 miles. Price at the factory was approx. $25,000. The fuselage framework was built up of spruce longerons, uprights and diagonals, with dural gussets at every joint; this framework was then painted with bitumastic paint and covered with "Alclad" aluminum sheet that was screwed to the framework. The wing panels were built up of laminated spruce spars and stamped-out "Alclad" ribs, after assembly the panels were fabric covered. The tail-group was built up of a mixed wood and metal construction, the horizontal surfaces were fabric covered, and the vertical surfaces were covered with "Alclad" sheet. The horizontal stabilizer was adjustable in in flight. Pilot's control column was of the "dep" (wheel) type; a 3-bladed metal propeller, electric inertia-type engine starter, and wheel brakes, were standard equipment. The cabin interior arrangement was optional, and some models of this type came out as very luxurious "air yachts". For this same airplane powered with a 9 cylinder Wright "Cyclone" engine of 500 h.p., see chapter for A.T.C. #67 in this volume.

Fig. 222. First Costa Rica to U. S. A. air-mail arrives in Loening "Amphibian" at Cristobal from San Jose on Dec. 28, 1928 via Pan American Airways.

A.T.C. #67
(8-28)
LOENING "AIR YACHT" (CYCLONE)

Fig. 223. A Loening "Air Yacht" with 500 h.p. "Cyclone" engine, riding anchor in the basin of a swanky "yacht club".

This model of the Loening "Air Yacht" was more or less identical to the model previously discussed in the chapter for A.T.C. #66; that is, it was identical in most all respects except for the powerplant installation which in this case was the 9 cylinder Wright "Cyclone" engine of 500 h.p. The performance with this installation was more or less the same as with the 500 h.p. Pratt & Whitney "Hornet" engine; although the "Cyclone" powered version proved later to have the edge and to be somewhat the more popular installa- in the "Air Yacht". We can assume that this was only a matter of preference.

The Loening "Air Yacht" in this version, was introduced and first shown about mid-year and it's unusual design and configuration was a source of interest to many; it was received quite well and very soon a number of them were sold, mostly for use as an executive-transport in the field of business. A popular model was a custom built type with a special interior and arrangement for the luxurious transport of 4 passengers in splendor and comfort. A standard model was also available to carry pilot and 5 or 6 passengers. A type certificate number for this version of the "Air Yacht" was issued in August of 1928 and it was manufactured by the Loening Aeronautical Engineering Co. in New York City. The factory was ideally situated with a ramp into the East River.

Listed below are specifications and performance data for the "Cyclone" powered Loening "Air Yacht" amphibian; span upper and lower 46'8", chord both 72", wing area 517 sq. ft., airfoil "Loening", length overall 34'8", height wheels down 13'2", height wheels up 11'5", empty wt. 3849, useful load 2051, payload 1013, gross wt. 5900 lb., max. speed 122, cruise 100, land 52, climb 850, ceiling 13,500 ft., gas cap. 140 gal., oil 12.5 gal., range approx. 500+ miles. Price at the factory was approx. $25,000. The fuselage framework was built up of spruce longerons, spruce uprights and diagonals, with "dural" gussets at every joint; this framework was then painted with bitumastic protective paint and covered with "Alclad" aluminum sheet that was screwed to the framework. The wing panels were built up of laminated spruce spars and stamped-out "Alclad" ribs, after assembly the panels were fabric covered. The tail-group was built up of a mixed wood and metal construction, the horizontal surfaces were fabric covered and the vertical surfaces were covered with "Alclad" sheet. "Alclad" by the way, was a sheet of duralumin alloy with a facing on both sides that was of pure aluminum, this was to make it corrosion resistant. The fuel tank was in the fuselage below the pilot's section, and the oil supply tank was in the engine nacelle. A 3-bladed metal propeller, electric inertia-type engine starter, and wheel brakes, were standard equipment. Custom interiors and arrangement were available on order. For the next development in the Loening "Air Yacht" series, see chapter for A.T.C. #90 and #91 in this volume. These chapters discuss the models C2C and C2H.

A.T.C. #68
(8-28)
CURTISS-ROBERTSON "OX-5 ROBIN"

Fig. 224. A later example of the Curtiss "Robin" as built by Curtiss-Robertson. A favorite for business-flying.

This version of the "OX-5 Robin" was the production model as built by Curtiss-Robertson at St. Louis (Anglum), Missouri; production on the "Robin" commenced about July of 1928. Typical in form, it was also a 3 place high wing cabin monoplane, powered with the OX-5 engine, and was basically the same as the earlier model developed and built by the Curtiss parent company; with the incorporation of some minor changes from time to time. Some of these models were built with the "airfoiled" lifting struts and shock-cord sprung landing gear as shown, and some were built with the streamlined wing struts of much thinner section, and later types had "oleo" type landing gear. All models had the swiveling tail-skid. The three place seating arrangement was still the same, that is, the pilot sat up front in a wicker or metal bucket-type seat and the two passengers sat side by side on a bench-type seat in back. Dual-stick control was available for student pilot instruction. The "OX-5 Robin" enjoyed a considerable amount of popularity among the small operators both for profit and pleasure, and soon became a very familiar figure around airports across the whole country during this period. The "Robin" was gentle and very well suited for flight training purposes and was used extensively by the "Curtiss" flying school which were being located all over the country, and soon became the largest flying school system in the world.

As a contrast to the average, the "OX-5 Robin" offered a comfort and broad utility not present in the general run of OX-5 powered biplanes, offering features that were surely to appeal to many of those in search of a personal type airplane; except of course, to those genuine "aviators" to whom flying was open cockpits with wind in the face, the unmuffled roar of a good running engine, the unhampered visibility in all directions and the significant "tune" always played by the bracing wires! These were inherent qualities of the open cockpit biplane that were still loved by thousands, and to some extent even unto this day. To many, the "Robin" lacked this "romance of flying", but in it's short time the "Robin" won over a good many "converts".

The "OX-5 Robin" was definately not a "sport type" airplane, but it did have pleasant flight characteristics and a very good performance for a ship of this type. Yes, the "Robin" was a sturdy and good flying airplane and performed quite well; with the struggling efforts of the average OX -5 it could "top out" at an easy 99 m.p.h., cruise about 84, and landed with big ship "feel" at 44 m.p.h. In view of the "Robin' s" popularity and of the large number that were built, it is not too surprising to hear that a good number of these are still flying; though most have been modified to mount the more modern engines.

The "Robin" was equipped with two 25 gallon fuel tanks, one in each wing root for a maximum gas capacity of 50 gal., this was good for a range of some 6 hours or over 500 miles. With a full load of three people aboard, especially if they were very large people, the maximum gas capacity was not used. The normal fuel load with 3 aboard, was 30 gallons. The type certificate number for this model of the "OX-5 Robin" was issued in August of 1928 and it's production continued until sometime in late 1930, when all practical supply of OX-5 engines was finally exhausted! Undaunted nevertheless, an effort was made to prolong the life of this series and some "Robins" were later built that were powered with the "Tank" engine, which was basically an "air-cooled" OX-5 engine.

Douglas Corrigan's "Wrong Way" Robin was originally one of this type and was powered with an OX-5 engine, but it was modified with the installation of a 5 cylinder Wright "Whirlwind" J6 engine of 165 h.p. for his "unplanned" venture across the Atlantic Ocean to Ireland! This famous "Robin" is still in existence, it has been carefully preserved and may someday be rebuilt to fly again!

Listed below are specifications and performance data for the OX-5 powered Curtiss-Robertson "Robin"; wing span 41', chord 72", wing area 224 sq. ft., airfoil "Curtiss C-72", length 25'10", height 7'10", empty wt. 1489, useful load 728, payload (with 30 gal. gas) 378, gross wt. 2217 lb., max. speed 99, cruise 84, land 44, climb 420, ceiling 10,200 ft., gas cap. (max.) 50 gal., gas cap. (normal) 30 gal., oil 4 gal., normal range (4 hrs.) 340 miles. Price at the factory held up around $4000, but plummeted to a low price of $2495 in 1930. The fuselage framework was built up of welded chrome-moly steel tubing, lightly faired to shape and fabric covered. The one-piece semi-cantilever wing panel was built up of solid spruce spars and "Alclad" aluminum stamped-out ribs, also fabric covered. The air-foiled struts were a main tube of chrome-moly steel with metal stamped-out ribs along it's length and were also fabric covered. The use of the "airfoiled" wing strut was more or less pioneered by "Bellanca" and found their way to use on many monoplanes during this period. The so-called "thin struts" on the later "Robins" were of streamlined chrome-moly steel tubing. The seats were built up of steel tubing and wicker, on the earlier models, but later were upholstered in various fabrics. The fabric covered tail-group was built up of welded chromemoly steel tubing and metal stamped-out ribs, the fin was ground adjustable and the horizontal stabilizer was adjustable in flight. A metal propeller and wheel brakes were available, the wings were wired for lights. Starting from early 1929, the "Robin" was available on the time-payment plan. For the "Challenger" powered version of this "Robin", see chapter for A.T.C. #69 in this volume.

Fig. 225. The "OX-5 Robin" was built in great numbers, a familiar sight at fields all over the country.

A.T.C. #69
(8-28)
CURTISS-ROBERTSON. "CHALLENGER ROBIN"

Fig. 226. Curtiss-Robertson "Robin" Model C with 6 cyl. Curtiss "Challenger" engine of 170 h.p. The "Robin" was a likeable airplane that sold in the hundreds.

This model shown here in various views, was the production version of the "Challenger Robin" as built by Curtiss-Robertson at St. Louis, Mo. It was a slight modification of the earlier Curtiss-developed "Challenger Robin" as described previously in the chapter for A.T.C. #63 in this volume. The production version was typical in most all respects except for a few minor changes that were incorporated from time to time; it was available with either the air-foiled wing struts or the streamlined steel tube struts, it now had "spring and oleo" struts incorporated into the landing gear and used the swiveling tail-skid thatwas rubber disc sprung. Wheel brakes were available and were optional. The earlier "Robins" used the old type of Curtiss-Reed "prop" that was bent from a "dural slab", but this version of the "Robin" used the new Curtiss-Reed propellerthat was a dural forging.

This latest "Robin" was also a 3 place high wing cabin monoplane and the seating was identical to the previous models, dual controls for student instruction were also available. The powerplant was the 6 cylinder Curtiss "Challenger" engine that was now approved and rated at 170 h.p. at 1800 r.p.m., and the engine cowling was now amply louvered for adequate ventilation; the exhaust collector-ring was buried in the engine cowling. There were two 25 gallon fuel tanks, one in each wing root; providing a maximum gas capacity of 50 gallons which was good for a cruising range of about 5 hours or 510 miles.

This model of the "Challenger Robin" was of a type similiar to that used by Jackson and O'Brine in setting an endurance record of over 420 hours during July of 1929, by refueling in flight. The re-fueling "tanker" was another "Robin" of this type. This record was shattered almost a year later by the Hunter Brothers in a Stinson monoplane, setting a record of over 553 hours during June of 1930. Not to be outdone, Jackson and O'Brine boarded another "Robin", the "Greater St. Louis", and recaptured their record during August of 1930 by staying up over 647 hours! About 5 years later another "Robin", named "Ole Miss", broke the existing record by staying up over 653 hours, and so it went.

The "Challenger Robin" was a sturdy combination with a fairly good performance and plenty of stamina that was proven beyond question on these many record flights, and in the many years of ordinary every-day use. A Model B "OX-5 powered Robin" that had serial #6 (NC-7145) built Aug. 28-1928, was still flying and rendering dependable service as late as 1953, although it had since been modified by the installation of a 220 h.p. Continental engine. And...it is known that there are many more "Robins" still flying at this time, giving their owners useful, dependable service.

The type certificate number for this model of the "Challenger Robin" was issued in August of 1928 and it was manufactured by the Curtiss-Robertson Airplane Mfg. Co. at St. Louis (Anglum), Mo. Ralph S. Damon was factory manager and later in 1929 became V.P., these were but stepping-stones in a skyrocketing career that brought him way up the ladder in the mammoth Curtiss organization. The next development in the "Robin" series were the "C" models.

Listed below are specifications and performance data for the Curtiss "Challenger" powered Curtiss-Robertson "Robin" model B; wing span 41', chord 72", wing area 224 sq. ft., airfoil "Curtiss C-72", air-foiled strut area 41 sq ft., length 24'1", height 7'10", empty weight 1576, useful load 864, payload 394, gross wt. 2440 lb., max. speed 118, cruise 102, land 47, climb 650, ceiling 12,700 ft., gas cap. 50 gal., oil 5 gal,. range 510 miles. Price at the factory averaged at $7,500. The fuselage framework was built up of welded chrome-moly steel tubing of square and round section, lightly faired to shape and fabric covered. The wing framework was built up of routed spruce spars and "Alclad" aluminum stamped-out ribs, also fabric covered. The "Robin" was one of the first airplanes to use metal stamped-out ribs in a production model, a feature that later became almost universal for commercial type aircraft. The air-foiled wing struts were built up of a round chrome-moly steel tube, using metal stamped-out ribs that were fastened along it's length and then fabric covered. The so-called "thin strut" version used wing struts of chrome-moly steel tubing that were streamlined in section. The "Robin" was not an exciting airplane but it was very likeable and many hundreds were sold in the few years that it was produced.

Fig. 227. This "Robin" had "Challenger" engine No. 4, an early version that was company ship for Standard Oil of New Jersey.

A.T.C. #70
(9-28)
"MONOCOUPE", MODEL 70

Fig. 228. 1928 "Velie" powered "Monocoupe" Model 70. This particular ship was flown by Phoebe Omlie in the National Air Tour for 1928.

Scattered among us in the hearts and minds of many, are even yet a countless number of fond memories and delightful stories of the pert little "Monocoupe". An airplane that had captured so many hearts with it's wholesome charm, and went on to later hit the veritable jack-pot in popularity. And, it really was no wonder, for this was a jolly and friendly sort of airplane that was absolutely contagious, and every owner-pilot was a willing and enthusiastic booster for the type. The "Monocoupe" owners and flyers were a happy lot, and were almost obnoxious in their enthusiasm and constant praise for this little ship. Truly, as Mono Aircraft had proudly announced, nearly 90% of all the light airplanes produced and sold in this country in 1928 were "Monocoupes".

The "Model 70" as pictured here, was an improved version to some extent but was still a good deal typical to the earlier models as built by "Central States". The sturdy axle-type landing gear was retained, as were the large characteristic side-windows which were somewhat outlandish, but identified the "Monocoupe" so quickly and easily. This new model "70", as put into production, was now powered with the newly introduced 5 cylinder "Velie" engine of 55 h.p., an engine that proved to be such a shot in the arm for the "Monocoupe" combination. Various engines were tried previously, but this one did the trick; it was at last a perfect mate for this little ship. The Velie engine was an "air-cooled radial" of 5 cylinders and was rated for 55 h.p. at 1850 r.p.m.; it was allowed 62 h.p. at 2000 r.p.m. for take-off. Well built and thoroughly reliable, the "Velie" soon gained a good reputation and began to be used even on aircraft of other makes. Due to the sudden and pyramiding success of the "Velie M-5", the company also experimented with a 9 cylinder air-cooled radial engine; this was the "Velie L-9" of 180 h.p. at 1900 r.p.m., but it proved somewhat unsatisfactory and was soon discontinued.

Quite anxious to show the little "Monocoupe" off in it's new dress, two of the "Model 70" were flown in the 1928 National Air Tour by Jack Atkinson and Phoebe Fairgrove Omlie, finishing in 19th and 24th places respectively. Not such a bad showing for a little ship, considering the rough terrain and the tough competition encountered on this 6300 mile "tour" which stopped at 32 cities. Many will recall Phoebe Omlie, the first woman to hold a "Transport Pilot" license in the U.S., as a gracious lady; a likeable and fitting "ambassadoress" to spread goodwill towards the "Monocoupe" and she managed a very fine job in this respect. Exploiting the merits of the little cabin monoplane as a

Fig. 229. The "Monocoupe 70" was the best selling light plane in the U.S. during 1928.

personal-type airplane that people would enjoy owning and flying.

The chumminess of side by side seating appealed to the private-owner flyer, and this also proved invaluable for pilot training; many flying schools across the country used "Monocoupes" in their primary course. The type certificate number for the "Velie" powered "Model 70" was issued in September of 1928; the model was designated after it's "A.T.C." number. The "Monocoupes" were now manufactured by the newly formed Mono Aircraft Corp. at Moline, Ill., which was still under the guidance of the capable energies of Don Luscombe. The early "Model 70" type was exhibited at the Detroit Aircraft Show in April of 1928, 3 ships were shown; one had the 6 cyl. "Anzani" engine, one had a 4 cyl. in-line British "Cirrus" engine, and one had a 5 cyl. "Siemens" engine. Some reports also list a "Velie" powered model as being on exhibit, but this has not been verified. The next development in the "Monocoupe" series was the "Velie" powered "Model 113", see chapter for A.T.C. #113.

Listed below are specifications and performance data for the "Velie" powered "Monocoupe" Model 70; wing span 32', chord 60", wing area 143 sq. ft., airfoil Clark Y, length 19'9", height 6'3", empty wt. 795, useful load 555, payload 215, gross wt. 1350 lb., max. speed 98, cruise 85, land 37, climb 550, ceiling 10,500 ft., gas cap. 25 gal., oil 2 gal., range approx. 500 miles. Price at the factory was approx. $2500. The fuselage framework was built up of welded steel tubing, faired to shape and fabric covered. The wing framework was built up of solid spruce spars and spruce and basswood built-up ribs, also fabric covered. The two fuel tanks were in the wing, one either side of the fuselage for a maximum capacity of 25 gallons. The wing struts were of chrome-moly steel in round section, with a light steel tube fairing framework attached that was fabric covered. The landing gear was built up in the same manner. The fabric covered tail-group was built up of welded steel tubing. This model was also tested as a float-seaplane with "Edo" floats. Visibility from the "Monocoupe" was excellent due to the over-size cabin windows, there was also a large sky-light overhead for visibility in that direction.

Fig. 230. "Monocoupe 70" in Canada (was NC-6554, c/n 81).

A.T.C. 71
(9-28)
"SPARTAN" (RYAN-SIEMENS), C3-1

Fig. 231. An early example of the 1928 "Spartan" biplane with "Ryan-Siemens" engine of 125 h.p.

In tracing our way back to an early and almost forgotten part of "Spartan" aircraft history, we come upon the scene of a typical modest beginning. A beginning that was prompted by the successful design and enthusiastic reception of a prototype airplane. Willis C. Brown, a practical man of purpose, diverse talents and good sound ability, felt compelled to join the number of others about the country that were busily engaged with newly awakened interest, in that never-ending search for the ideal light commercial airplane. With convictions already attained thru' experience and a good number of years of critical study, requirements for an airplane of this type were self evident and these were incorporated into the original design. So then, it surely would not be out of place to say that Willis Brown had a sincere conviction that his worthy contribution towards this end, might very well be the answer.

Introduced early in 1927, this initial effort put into production under the label of the Mid-Continent Aircraft Co., was a 3 place open cockpit biplane of typical "Spartan" lines as we later came to know them, and it was built by a handful of craftsmen and enthusiasts gathered together at Tulsa, Okla. The first production airplane and the many that followed thereafter, were all powered with the "Ryan-Siemens" (Siemens-Halske) engine; a popular 9 cylinder air-cooled "radial" engine of German manufacture that was rated at 125 h.p.

Aviation in all it's forms was very much in favor at this particular time, and investors looking for a good thing were more than eager to provide financial backing, and to take an active part in an industry that was beginning to show great promise. W. G. Skelly, a prominent and wealthy Oklahoma oil-man, and a few other air-minded business men, became interested in Brown's attractive looking project and purchased the assets of the Mid-Continent Aircraft Co. in the fall of 1927. The Spartan Aircraft Co. was soon after organized with Willis Brown as it's president; A. K. Longren, himself an engineer and former airplane manufacturer, was in charge of production. Actual incorporation of the firm was Jan. 17-1928. The general outlook was indeed great and they at "Spartan" were quite confident and thoroughly impressed with the possibilities in view for airplane manufacture at this time. Plans for increased production of the "Siemens" powered "Spartan" biplane were tended to immediately and work was also begun on a new enlarged plant, being put into operation about August of 1928. This was a plant with much better facilities and a much larger capacity, conveniently located next to a rail-siding and just a short way from the new Tulsa airport where the assembly plant, hangars, service station, and sales office were located.

The "Siemens-Spartan" model C3-1, was their first production type to receive it's approval for a type certificate number, which was issued in Sept. of 1928. The "Spartan" biplane was one of the few airplanes that had went through a complete stress analysis long before it came up for certification, so it got thru' this phase of it's tests in jig-time and flying colors! By this time, the "Spartan" had already been through an extensive series of shake-down tests, and a few had already been built and sold; it had also been shown at the Detroit Aircraft Show of April in 1928. A good amount of interest was shown for this version, which quickly established itself as a neat and well-mannered airplane. Good manners was a built-in trait of the design that was to be more or less characteristic of all the numerous "Spartan" biplanes that followed in succession. As pictured here in various views, the "Spartan" was an open cockpit biplane of fairly average configuration, but with a definite stance and a modest air of quality all it's own. The 9 cylinder "Ryan-Siemens" engine of 125 h.p. combined very nicely with the "Spartan" design to give a smooth andseemingly effortless performance.

Memory can still recall when quite by chance we saw our first "Spartan", it was a demonstrator that had dropped in on the local field to drum up a little interest for the type. (During this period of aviation growth, it seemed that just about everyone had "wares" to sell, and the demonstration tour was a custom that was a source of endless interest to those that flew or just hung around, and was always something to look forward to). From more than just casual observation, it was soon evident that everyone present that morning was thoroughly impressed with the effortless way the "Spartan" flew, especially in and out of the small field, and most seemed to be more than a little intrigued with the "Siemens" engine; it was running exceptionally well and running very smooth.

This particular "Spartan" type was indeed a pleasant flying airplane and was quite ideal for general all-purpose flying; with it's docile behavior and compatible nature, it was especially a standout as a training plane; this provided a good basis for operations, and prompted the formation of the famous "Spartan School of Aeronautics". The model C3-1 biplane was somewhat slow in pickingup sales, for a while, but eventually it did build up a staunch following, and a fair amount of country-wide popularity. Nearly 100 of this model were built before the "Siemens" engine became almost unavailable due to some labor strife in Germany. Looking hurriedly about for a replacement powerplant in this power range, Willis Brown had not much to choose from actually, so the 4 cylinder "Caminez" engine of 120 h.p. had already been tried out. The airplane performed quite well with this combination, but the "Caminez" engine was still plagued with many development problems and was not entirely satisfactory. Only a prototype of this version was tested. A new version was introduced shortly after that was

Fig. 232. The first "Spartan" biplane was flown by Willis Brown on it's maiden flight Oct. 25-1926. Willis Brown on right, Paul Meng on left.

powered with the 9 cylinder "Walter" (Czech-slovakian) engine of 120-135 h.p., this became the model C3-2; see chapter for A.T.C. #73 in this volume.

Willis Clinton Brown, the designer of the "Spartan" biplane, was a former Air Service pilot whose active participation in aviation and airplane manufacture actually dated back to 1912, when as a boy of 16 he had built his first airplane. Some years after W.W. I, he had owned and flown "Waco" 7 and 9 type airplanes among others, as a flying-salesman out of the Tulsa area during 1924-25. His experiences as a flying-salesman in this territory prompted him to design his first practical commercial-type airplane, an airplane which was calculated to operate efficiently and with a minimum amount of piloting technique under the varied and difficult conditions most always encountered in this type of flying. The prototype, as shown here, was designed thru 1925 and it was built in a vacant mattress factory. It was finally test-hopped by Willis Brown himself on October 25-1926. This airplane was a 3 place open cockpit biplane and was powered with a LeRhone "rotary" engine of 80 h.p. which had been re-worked and converted into a static-radial engine, a conversion called the "Super Rhone". Performance of this engine was often unpredicatble, but even at that, the performance and flight characteristics of this first airplane were so satisfactory that Brown was persuaded to consider building replicas of the type; this eventually led to the formation of the Mid-Continent Aircraft Co. at Tulsa, Oklahoma.

Listed below are specifications and performance data for the "Ryan-Siemens" powered "Spartan" biplane model C3-1; span upper and lower 32', chord both 60", wing area 291 sq. ft., airfoil Clark Y, length 23'6", height 8'9", empty wt. 1355, useful load 800, payload 370, gross wt. 2155 lb., max. speed 115, cruise 98, land 42, climb 720, ceiling 11,000 ft., gas cap. 44 gal., oil 4.5 gal., range 450-500 miles. Price at the factory was about $5,200. The fuselage framework was built up of welded chrome-moly steel tubing, deeply faired to shape with wood fairing strips and fabric covered. The wing panels were first built up of spruce and plywood box-beam spars, but this was later changed to solid spruce spars that were routed to an "I beam" section, ribs were built up of spruce and plywood and the panels were fabric covered. All interplane struts were of chrome-moly steel tubing in a streamlined section, interplane bracing was of streamlined steel wire. The fabric covered tail-group was built up of spruce box-beam spars and spruce and plywood built-up ribs in a heavy section that provided exceptional rigidity, the movable surfaces were all cable operated. The landing gear was light but robust and was built up of chrome-moly steel tubing in both round and streamlined section, it was of the split cross-axle type using two spools of wound rubber shock-cord for snubbing the taxi and landing loads. The tail-skid on the early models was a steel tube structure that swiveled and was rubber shock-cord sprung, but this was later changed to the steel spring-leaf type with a hardened shoe. The fuel tank was in the center-section of the upper wing and could be removed without disturbing the rigging, the baggage compartment was under the spacious front seat. The standard color scheme for the "Spartan" biplanes was of a rich maroon for the fuselage and tail-group, with bright orange-yellow wings.

Fig. 233. 1928 "Spartan" C3-1, with Ryan-Siemens engine. The reputation of the "Spartan School of Aeronautics" was based on the docile behavior and compatible nature of this airplane.

A.T.C. #72
(9-28)
CESSNA - MODEL AW

Fig. 234. The "Cessna" Model AW, powered with Warner "Scarab" engine of 110 h.p.

This long-time favorite "Cessna" was a sensitive little beauty possessed of very feminine traits and a completely infectious quality; in a word, it was a charming little airplane. It seemed to be such a natural and compatible combination with the 7 cylinder Warner "Scarab" engine of 110 h.p. With this spunky little air-cooled radial engine coupled to the sensible and clean design of the "AW" the performance derived turned out to be just a little short of remarkable. Remarkable that is, in the sense that it was a 4 place airplane that offered performance usually enjoyed only with higher powered aircraft, and all this was made possible by the sensible use of 110 h.p. During this period of aviation development it became almost imperative for an airplane "type" to prove itself worthy before one and all, by making a good show in the many air-derbies and air-races that were being held frequently in various parts of the country. "Cessna's" numerous rousing victories against airplanes packing much more horse-power, proved it to be an airplane of good breed and very high caliber. It certainly proved it's sound design and proved also that speed was one of the "Cessna's" foremost inherent qualities. Matched power for power, it would out-perform many of the best of them.

In it'sbasic form the Model AW as pictured here, was still typical to the earlier "Anzani" powered Model AA; it was a fully enclosed high wing monoplane with a stout one-piece full cantilever wing and a slim, trim fuselage that seated four. Perhaps a little tight, but chummy. Two passengers were seated in the back on a bench-type seat and the pilot and one passenger sat up in the forward section, using the front spar as a head-rest. Visibility was quite good due to the absence of struts and wires, and two handy doors were provided for easy entrance and exit. The Cessna AW was often described as very feminine and capricious, but when mastered with a firm hand and good understanding, it behaved quite well; being capable of a performance that was scintilating yet economical and efficient. These little monoplanes, if coaxed and treated right, were very well-behaved and though nimble and spirited, were still inherently stable from most attitudes through the "pendulum stability" of the high wing design. This basic design was so satisfactory that it was used without much change for over 25 years. Clyde V. Cessna had always been an enthusiastic advocate of the monoplane type, because of it's greater aerodynamic efficiency, and especially for the monoplane of the full cantilever internally braced type; his convictions have since been proven beyond a doubt.

For two years running, Earl Rowland and

Fig. 235. Though capricious and very feminine in behavior, the "Cessna AW" was an outstanding airplane.

his Model AW pictured here, were beating everything in class and often out of class. This airplane was used so successfully in the 1928 air-derby and air-race season, and the only changes made to the airplane for the 1929 seasonwere the addition of an "N.A.C.A. type" anti-drag cowling over the Warner engine. Earl Rowland was first in the Class A division of the 1928 Air Derby from New York to Los Angeles, during the air-races at L.A., he came in first in two civilian free-for-all heats. He placed 13th in the National Air Tour for 1929, then won Class D in the Miami to Cleveland Air Derby. Another "Cessna AW", also shown here, was flown by Parker "Shorty" Cramer to a new record for light planes in early 1929 by making a flight of more than 10,000 miles from Wichita, Kansas to Detroit, Mich. and then on to Nome, Alaska and into Siberia, with a return trip ending in New York City. It was the first commercial airplane to make such a trip into frozen Siberia, and this was surely a phenomonal flight for such a light airplane. But the Model AW was not always dashing around breaking records, it did serve and very well too, as a personal-type airplane. Business houses used it to transport salesmen and executives around the country, and it was even known to carry passengers up on "Sunday rides" at "three bucks" apiece.

The type certificate number for the "Cessna AW" was issued in September of 1928 and the model was built through 1929, according to factory records 50 airplanes of this type were built. The Model AW was quite popular and was going very good, with prospects ahead for a great future but when the bottom dropped out of the aircraft market due to the crippling "depression", Cessna found it rather hard to keep going and the "AW series" production was halted for the time being. It was still built on order, and then the "DC-6 series" were developed as sort of a last-ditch stand.

Just a few years ago there was a forlorn looking "Cessna AW" parked off to the side at one of the airports, and inspection revealed that it was absolutely rotting away; it was

Fig. 236. Earl Rowland and the "Cessna AW" had chalked up many rousing victories in the 1928-29 season. Note excellent view of Warner "Scarab" engine.

such a pity, because it was already beyond practical repair and probably would never be rebuilt. But leave us not dwell on this sombre note, because there are still a few of the old "Model AW" flying. The "Cessna" Model AW was manufactured by the Cessna Aircraft Co. at Wichita, Kansas with Clyde V. Cessna as the president and chief engineer.

Listed below are specifications and performance data for the Warner "Scarab" powered "Cessna" Model AW; wing span 40', chord at root 86", chord at tip 58", (M.A.C. 66"), wing area 224 sq. ft., airfoil "Cessna" (Mod. M-12), length 24'9", height 6'11", empty weight 1225, useful load 1035, payload 550, gross wt. 2260 lb., max. speed 125+, cruise 105, land 42, climb 620, ceiling 12,000 ft., gas cap. 40 gal., oil 4.5 gal., range 630 miles. Price at the factory was $6,900 with a wood propeller, and $7,115 with a metal "prop", this was later raised to $7,500. The fuselage framework was built up of welded chrome-moly steel tubing, faired to shape with wood fairing strips and fabric covered. The wing framework was built up of one-piece laminated spruce spars and spruce and mahagony plywood built-up ribs, also fabric covered. The internally braced full cantilever wing was built up in one continuous piece and was tapered in plan-form and section, it was bolted directly on top of the fuselage. The forward occupants used the front spar of the wing as a head-rest, the two rear occupants were seated between the spars with head-room provided by the omission of ribs in that area. For other construction details refer to previous chapter covering the model AA.

Fig. 237. This "Cessna AW" had flown into Siberia and back, a trail-blazing trip of over 10,000 miles.

A.T.C. #73
(10-28)
"SPARTAN" (WALTER), C3-2

Fig. 238. Spartan C3-2 with 120 h.p. Walter engine.

This next development of the "Spartan" biplane was comparatively similiar to the first version that was powered with the Ryan-Siemens except for the change in engine installations, which in this case was the very interesting 9 cylinder "Walter" engine of 120-135 h.p. This "air-cooled radial" engine was of Czechoslovakian design and manufacture, and had already built up quite an enviable reputation abroad for it's performance and exceptional reliability. Spartan Aircraft chose this engine as their standard powerplant installation for the Model C3 when "Siemens" engines became unavailable, and in September of 1928 they announced their exclusive rights for the distribution of the "Walter" engine in this country. But this engine actually failed to stir up very much interest nor sell in any great number, because there was always the hope that good American "radials" were finally coming along and naturally they would be preferred. Spartan sold a few of the "Walter" engines to "Command-Aire" to power one version of their 3 place biplane, and a few other companies placed trial orders sparingly.

In it's basic form, this model of the "Spartan" was also a 3 place open cockpit biplane with wings of equal span in a single bay, and having no interplane stagger. The robust landing gear was also of the split cross-axle type using wound rubber shock-cord to snub the bumps, and the tail-skid was now of the steel spring-leaf type. The cockpits were deep and roomy, offering comfort and good protection against the weather. Typical in every sense, the performance of this "Walter" powered version was very good in all respects and though it was quite spry when need be, it was also a well-mannered airplane. Of course this remained a true characteristic of all the "Spartan" biplanes. This "Spartan" biplane being also of a rather modest and quiet personality failed to excite the flying public to any tremendous extent, but being ideally situated in the proper locale, they did sell a good number of these airplanes to the "oil men" and related concerns for personal and company use. The type certificate number for the model C3-2 was issued in October of 1928 and then reissued for some modifications as the C3-120 in October of 1929.

A "Walter" powered "Spartan", shown here, came into some prominence for a time when it was flown non-stop by the very capable Leonard S. Flo on a flight from Walkersville, Ontario to Key West in Florida in Nov. of 1928. Covering the distance across the

Fig. 239. Spartan C3-2 (C3-120) powered with 9 cyl. "Walter" engine. Used by Spartan School of Aeronautics for pilot training.

entire length of the U.S.A. in 17.5 hours. Proving it's exceptional ability, the airplane took off with a load that was the equivalent to at least 120% of it's empty weight, or a useful load of about 1380 lbs! Despite the demonstrated and very apparent usefulness of this combination, it still failed to sell to any great degree so the Spartan organization began experimenting with other power installations. There was a C3 version that was powered with the 6 cylinder Curtiss "Challenger" engine of 170 h.p. that worked out fairly well, and another version that was powered with the new 7 cylinder "Axelson" engine of 115-150 h.p.; but neither proved entirely satisfactory so these developments were dropped. Finally, the new Wright "Whirlwind" J6-5 engine of 150-165 h.p. was tried in another version; this worked out very well and in fact it proved so satisfactory that it was finally adopted as the standard engine installation for the "C3 series" for the coming season. This model was first the C3-5, but when it received it's approval for a type certificate number it was redesignated the model C3-165. See chapter for A.T.C. #195 which will follow.

In 1928 Willis Brown was visiting in Europe to study their activities and also to acquire a suitable engine for the "Spartan", during his absence from this country the crew left at Spartan Aircraft had made unwarranted changes in the design of the "Spartan" biplane. This naturally had caused some

Fig. 240. Leonard S. Flo and Willis C. Brown.

enmity and caused a breech in relations, tactics that convinced Brown to sell his interests and pull out. When Brown severed connections with Spartan Aircraft, he made a study of the trend shaping up in this country and later went to "Warner Engines" as V.P. in charge of sales.

Listed below are specifications and performance data for the "Walter" powered "Spartan" biplane model C3-2; span upper and lower 32', chord both 60", wing area 291 sq. ft., airfoil Clark Y, length 23'6", height 8'8", empty wt. 1310, useful load 840, payload 380, gross wt. 2150 lb., max. speed 115, cruise 98, land 45, climb 720, ceiling 11,000 ft., gas cap. 49 gal., oil 4.5 gal., range 500 miles. Price at the factory was $5,250. See chapter #71 for construction details of the fuselage, wings, interplane struts and bracing, tail-group, landing gear, and tail-skid. The fuel supply was carried in a 49 gallon tank built into the center-section of the upper wing panel. This tank could be removed for servicing without disturbing any part of the rigging or airframe. The standard color scheme was a deep rich maroon for the fuselage and tail-group, with bright orange-yellow wings. These "Spartans" were built in Tulsa, Okla. by the Spartan Aircraft Co., with W. G. Skelly as the Chairman of the Board and Willis C. Brown as it's president and chief engineer.

A.T.C. #74
(10-28)
STINSON "DETROITER", SM-1DA

Fig. 241. "Eddie" Stinson, shown here, reflects pride in his "Detroiter" monoplane. This ship, an SM-1DA was the first of the SM-1D type with Wright J5 engine. Flown to 5th place in National Air Tour.

The colorful "SM-1 series" of the Stinson monoplane were known the world over and had been in production for over a year by this time in the Northville, Michigan plant; this model shown here was the improved version of this famous "Detroiter" series. A 6 place high wing cabin monoplane of rather large and buxom proportions that was powered with the time-tested 9 cyl. Wright "Whirlwind" J5 engine of 220 h.p. This latest development was more or less typical to the previous models (SM-1 & SM-1B) except for many minor improvements of the landing gear, brake system, engine cowling, cabin interior, and the like. Improvements that were either found necessary or desirable, but do not plainly show just by casual observance. A "Detroiter" of this type (SM-1DA) was flown by "Eddie" Stinson in the National Air Tour for 1928 and finished in 5th place; with the newly developed "Junior Detroiter" (see A.T.C. #48 in this volume) taking a 3rd, and another "Detroiter" a 6th place. "Stinson" was well represented in this reliability tour and did extremely well despite the determined competition encountered.

The "Detroiter" monoplanes in this latest series (SM-1D) were large and bulky looking, carrying 6 people and their baggage very comfortably; this all accounted for a gross weight of some 4500 pounds. In spite of this apparent bulk and seeming overload, the performance of this ship was admirable. New air-lines were beginning to blossom out in all parts of the country, new routes that carried both mail and passengers; it is only logical that many of these operated the Stinson "Detroiter" monoplane. The Tanner Motor Livery of Los Angeles, Calif. had an SM-1DA type (NC-9601) that they used occasionally on charter and sight-seeing trips in that area. The type certificate number for the model SM-1DA was issued in October of 1928; they were manufactured by the Stinson Aircraft Co. at Northville, Michigan. The genial "Eddie" Stinson was Pres., Wm. C. Naylor was chief engineer, and Randolph Page was pilot in charge of test and development.

There were only a mere handful of the SM-1DA type built, possibly 5, and most of these were operated with the J5 engine only for a short time. As an example: SM-1DA type registered NC-9600, operated a short while with a "Whirlwind" J5 engine as a 4 place passenger and mail carrier with Thompson Aeronautical Corp.; it was converted to the

SM-1D300 type by Stinson Aircraft with the installation of a Wright J6-9-300 engine; as the SM-1DX this same ship was powered by the new Packard Diesel engine and operated in test by T.A.C.; sometime later it was converted back to the SM-1D300 type with the installation of a J6-9-300 engine and operated in aerial photography and exploration. SM-1DA type registered NX-9617, was also converted to the SM-1D300 type and operated in this combination at least into 1931.

Listed below are specifications and performance data for the "Whirlwind" J5 powered Stinson "Detroiter" model SM-1DA; wing span 45'10", chord 84", wing area 280 sq. ft., airfoil "Stinson" (modified M-6), length overall 32', height 8'6", wheel tread 108", empty wt. 2432, useful load 2068, payload 1300, gross wt. 4500 lbs., max. speed 120, cruise 105, land 60, climb 700, ceiling 12,000 ft., gas cap. 90 gal., oil 6 gal., range 750 miles. Price at the factory field was $12,500. The fuselage framework was built up of welded chrome-moly steel tubing with steel gussets welded into every joint; fuselage was heavily faired to shape with plywood and metal formers and spruce fairing strips, then was fabric covered. The wing framework was built up of solid spruce spar beams and spruce and plywood built-up ribs, also fabric covered. There were two removable fuel tanks, one in each wing root flanking the fuselage. The fabric covered tail-group was built up of welded chrome-moly sheet and steel tubing, the fin was ground adjustable and the horizontal stabilizer was adjustable in flight. Dual control wheels of the "Dep" type, wheel brakes, inertia-type engine starter, metal propeller, and wiring for lights, was standard equipment. The standard color scheme was a deep rich maroon for the whole airplane with accenting striping; "Berryloid" finishes were used throughout. For the next development in this "Detroiter" series, the SM-1DB, see discussion for A.T.C. #76 in this volume.

A.T.C. #75
(10-28)
FAIRCHILD, FC-2C (CHALLENGER)

Fig. 242. Fairchild FC-2 with 6 cyl. Curtiss "Challenger" engine of 170 h.p. NC-32 was No. 3 (also as No. 2) first had Curtiss C-6. Taken Aug. 22, 1928.

The Fairchild FC-2C with Curtiss "Challenger" engine was somewhat of a rare-bird in the Fairchild monoplane series, and as far as can be determined, there were only a small number built and these were built for use by the Curtiss Flying Service. Basically, it was a 5 place high wing cabin monoplane similiar to the early model of the standard Fairchild FC-2 in most all respects; this version was of the early "pinch back" or "razor back" type, the only noticeable change was in the engine installation. This development of the FC-2 was powered with the recently introduced 6 cylinder Curtiss "Challenger" engine of 165-170 h.p., an engine of later proven stamina and capabilities that was enthusiastically championed by the Curtiss Motor division.

This airplane was used to good advantage by the Curtiss Flying Service as an air-taxi, also in their flying schools for certain advanced phases of flight instruction and doubled in fact as a test-bed for further development of the Curtiss "Challenger" engine. This version was probably a fairly good enough and even compatible combination, but it is doubtful if it was very sprightly as a 5 place airplane with this amount of power. The type certificate number for the Fairchild FC-2C (Challenger) was issued in October of 1928. For another FC-2C version that was powered with the 6 cylinder Curtiss in-line C-6 engine of 160 h.p., see chapter for Grp. 2 approval numbered 2-40 which will follow. Manufactured by the Fairchild Airplane Mfg. Corp. at Farmingdale, Long Island, N. Y.

Listed below are specifications and performance data for the Curtiss "Challenger" powered Fairchild monoplane model FC-2C; wing span 44', chord 84", wing area 275 sq. ft., airfoil Gottingen Mod., length 31'6", height 9', empty wt. 2239, useful load 1361, payload 740, gross wt. 3600 lb., max. speed 112, cruise 96, land 49, climb 540, ceiling 10,000 ft., gas cap. 75 gal., approx. range 750 miles. The fuselage framework was built up of welded steel tubing, lightly faired to shape with wood fairing strips and fabric covered. The wing framework was built up of spruce box-type spars and spruce and plywood built-up ribs, also fabric covered. This model also featured folding wing panels for space-saving storage. There were two fuel tanks, one in each wing root. The fabric covered tail-group was built up of welded steel tubing, the fin was ground adjustable and the horizontal stabilizer was adjustable in flight. Wheel brakes and a Curtiss-Reed metal propeller were standard equipment, the wings were wired for lights. The next development in the "Fairchild" monoplane series was the Model 71, see chapter for A.T.C. #89 in this volume.

A.T.C. #76
(10-28)
STINSON "DETROITER", SM-1DB

Fig. 243. Eddie Stinson flew this "Detroiter" to 5th place in the 1928 National Air Tour, shown here at Mills Field in San Francisco. Ship shown was first of the SM-1D type. SM-1DB was typical.

This model of the Stinson "Detroiter" monoplane discussed here, was another version in the latest "SM-1D series"; also a 6 place high wing cabin monoplane that was powered with the latest series of the venerable Wright "Whirlwind" J5 engine of 220 h.p. Exact differences between the various models in the SM-1D series monoplanes has been almost impossible to determine because of scant records available and because those craftsmen that were familiar with these ships do not seem to recall. Then too, there were only a small number of this series built; possibly only one SM-1DB type (NC-6580) which was manufacturers serial #D-301 and operated in the Detroit area for a few years. The model SM-1DB was issued this certificate in October of 1928; there was some variation in it's weights over the model SM-1DA. Empty weight of the SM-1DB was 90 pounds greater and this would tend to indicate the addition of some equipment or appointments that were lacking on the SM-1DA type.

Ed Stinson, beside being a master aviator, was his own star salesman. His frequent trips about the country brought about a number of sales and developed staunch supporters; many business men called upon were convinced that the airplane would be a great asset to their operations, and especially the spacious and comfortable Stinson "Detroiter" monoplane. The Stinson plant at Northville, Mich. had been developed to it's fullest extent by now and was bustling with the concerted effort of some 200 employees with a production capability that often reached 10 airplanes per week. Both the "Detroiter" series and the new "Junior Detroiter" series were being built in the various models; a good many were shipped abroad to near-by and far-away countries.

Specifications and performance data were more or less typical for all ships in this SM-1D series; wing span 45'10", wing chord 84", wing area 280 sq. ft., airfoil "Stinson" (modified M-6"), length overall 32', height 8'6", wheel tread 108", empty wt. 2522, useful load 1978, payload 1218, gross wt. 4500 lbs., max. speed 120, cruising speed 105, land 60, climb 700 ft. first minute, service ceiling 12,000 ft., gas cap. 90 gals., oil 6 gals., range 750 miles. Price at the factory field was about $12,500. The fuselage framework was built up of welded chrome-moly steel tubing, with steel gusset plates welded into every joint; framework was heavily faired to shape with plywood and metal formers and spruce fairing strips, then was fabric covered except for a portion of the cabin section. The semi-cantilever wing framework was built up of solid spruce spars and spruce and plywood built-up ribs, also fabric covered. There were two fuel tanks, one in each wing root flanking the fuselage. Dual control wheels of the "Dep" (Depperdussin) type, individual wheel brakes, parking brake, inertia-type engine starter, metal propeller, and wiring for lights was standard equipment. The cabin interior on these latest "Detroiters" was decked in tasteful appointments with the use of insulation and sound-proofing for comfortable and practically noise-free flying; flying that was just about as deluxe as one could get for this day and age. Photographic examples of the SM-1D series are practically non-existent but some apparent differences in the models were noted; the engine nose-cowling in some cases was rounded off and louvered, it did not use the long-familiar "spinner" enclosing the propeller hub. This version was probably the handsomest of the "Whirlwind J5" powered "Detroiter" monoplane series.

A.T.C. #77
(10-28)
STINSON "DETROITER", SM-1DC

Fig. 244. Craft shown is not a model SM-1DC, but was typical in most respects.

In March of 1928, Edw. Stinson and Geo. Haldeman flew a "Detroiter" monoplane over Jacksonville, Florida for 53 hours and 36 minutes to break the existing non-refueling endurance record. In August of 1928, Parker "Shorty" Cramer and Bert "Fish" Hassell took off from Rockford, Ill. in their "Detroiter" monoplane, the "Greater Rockford", for a flight across the top of the world to Denmark; they were in search of a better and shorter route across the Atlantic Ocean to Europe. Other "Detroiters" were also record bent; Byron Newcombe and Roy Mitchell flew their "City of Cleveland" to a new refueling endurance record of 174 hours plus; their refueling "tanker" was another "Detroiter" monoplane. Soon after, there were other endurance attempts at Buffalo, N.Y. and also at Houston, Texas. But, not all "Detroiters" were out breaking records, many of them worked long and hard for their keep. As a typical example; a Stinson SM-1DC (NC-9618, #D-307) was busily engaged as a "fish hauler". This particular ship, licensed as a single-place transport, hauled fish and other products of the briny deep on southern U.S. and Mexican routes; hauling a payload of some 1300 pounds. The operator of this craft was the International Air Transport operating out of Brownsville, Texas.

The Stinson "Detroiter" model SM-1DC was basically a 2 place craft and it's type certificate number was issued in October of 1928. Basically typical of the SM-1 series, the SM-1DC was also fitted with the familiar airfoiled wing bracing struts that contributed measurably to the total lifting area and the exaggerated dihedral angle greatly improved the lateral stability. The cabin interior of the SM-1DC was void of extra seats, cabin floor and walls were fitted to handle all manner of cargo.

Listed below are specifications and performance data of the J5 powered Stinson "Detroiter" model SM-1DC; wing span 45'10", chord 84", wing area 280 sq. ft., airfoil "Stinson" (modified M-6), length overall 32', height 8'6", wheel tread 108", empty wt. 2514, useful load 1986, payload 1226, gross wt. 4500 lbs., max. speed 120, cruising speed 105, land 60, climb first minute 700 ft., service ceiling 12,000 ft., gas cap. 90 gal., oil 6 gal., range 750 miles. Price at the factory field averaged around $12,500. Construction details of the model SM-1DC were typical of the series; see previous chapters discussing the

SM-1D type. It would be of interest to note that the most popular and most plentiful of the "Detroiter" monoplane series was the SM-1B type which was also certificated in October of 1928 under a Group 2 approval numbered 2-24; this was as a 6 place transport with a Wright "Whirlwind" J5 engine and an allowable gross weight of 3485 lbs. Performance of this earlier series was a good deal more sprightly than that of the heavily loaded SM-1D type. In all, some 74 of the SM-1 and SM-1B type were registered; the first 45 aircraft were eligible under A.T.C. #16 and the following 29 of them were certificated under Grp. 2-24 approval. The Grp. 2 approval was later issued to embrace all of the SM-1 and SM-1B type from serial number M-200 to M-273; other modifications to various ships in the SM-1 series are reflected in the numerous Group 2 approvals issued from time to time. These were special purpose approvals and often issued to only one airplane.

A.T.C. #78
(10-28)
STINSON "DETROITER", SM-1DD

Fig. 245. Outfitted for winter flying is one of the last examples of the J5 powered "Detroiter", the SM-1DD was typical.

For an airplane "type" that was so popular and so well known, it is saddening that so much data and a great many facts on the Stinson "Detroiter" are seemingly shrouded in great mystery. This is especially true of the 1928 Stinson SM-1D series which were the last of the "Whirlwind" J5 powered "Detroiter" monoplanes in regular production. All of the models in this SM-1D series were more or less typical, except for seating arrangement and some modifications brought about by use in different spheres of service. Of the four models in this series, and certificated practically in a group, all were of the same gross weight, but they did vary to a degree in their empty weights. The model SM-1DD was the lightest one of this group and this would indicate that it was more than likely outfitted with a rather bare cabin interior for hauling cargo. This airplane was listed as a "two place", indicating further that the cabin area was minus of seats to provide cargo space; a spacious compartment that carried over 1300 pounds of payload. The SM-1DD was some 242 pounds lighter than others in this series, having an allowable useful load of 2220 pounds which was almost equal to it's empty weight. The SM-1DD could have been profitably employed in "bush flying" operations, both on wheels and on skis.

"Eddie" Stinson never spent much of his time behind a desk, he spent much more of his time flying about the country, visiting Stinson dealers to keep tab on various plans and requirements, and studying and discussing possible market requirements for the times just ahead. Late in 1928, Ed Stinson foresaw that better performance and greater utility was becoming imperative and began to plan his strategy along two lines, either to bolster the 6 place "Detroiter's" performance with more horsepower to help it retain it's competitive advantage, or to concentrate more on the 4 place "Junior" models which were of ideal capacity for most business requirements and delivered a very satisfying performance in the higher powered versions. These plans led to the development of such types as the SM-1F (A.T.C. #136,) the SM-6B (A.T.C. # 217), the SM-2AB (A.T.C. #161), and the SM-2AC (A.T.C. #194); along with these developments there were variants of the SM-2 and the experimental SM-4 and SM-5. All of this accelerated development, plus the hub-bub of moving plant operations to a new site in Wayne, Mich., which would have larger capacity and an adjoining airfield, cut into actual production at this time and this would very well account for the scarcity of the SM-1D type.

Listed below are specifications and performance data for the "Whirlwind" J5 powered Stinson "Detroiter" model SM-1DD; wing span 45'10", chord 84", wing area 280 sq. ft. airfoil "Stinson" (modified M-6), length overall 32', height 8'6", wheel tread 108", empty wt. 2280, useful load 2220, payload 1320, gross wt. 4500 lbs., max. speed 120, cruise 105, land 60, climb 700, service ceiling 12,000 ft., gas cap. 90 gal., oil 6 gal., range approx. 750 miles. Price at the factory field averaged $12,500. Construction details of the SM-1D were typical of others in the SM-1D series, see previous discussions. The fin was ground adjustable and the horizontal stabilizer was adjustable in flight; wheel brakes, inertia-type engine starter, metal propeller, and wiring for lights, was standard equipment. Later developments in the SM-1D series were the Wright J6-9-300 powered SM-1D300 which was under Grp. 2-60 approval and the Packard "Diesel" powered SM-1DX which was under Grp. 2-228 approval.

A.T.C. #79
(11-28)
CONSOLIDATED "TRUSTY", PT-1

Fig. 246. Consolidated PT-1 with Wright "E" (Hispano-Suiza) engine. Void of frills, with simple lines and rugged structure.

The Consolidated PT-1 was probably one of the most famous and well-remembered primary trainers of all times, and for a good period of time was used exclusively by the "Air Corps" as a primary-trainer; used to give their "kaydets" their initial baptism in the manly art of military flying. It has been generally agreed that the advent of the PT-1 in early 1925 was heralded as the passing of the venerable "JN-4D", or better known as the "Jenny", that had been used by the Air Service as a training ship for so many years.

A glimpse at the accompanying views will clearly bear out the fact that the PT-1 was a lean, stark, and rather ugly airplane; surely no thing of beauty in the general sense but none will deny that it was just about perfect for the job intended. They were quite amiable and extremely rugged in nature, and it took a pretty sloppy "dodo" using extra effort to hurt one of these airplanes! In it's basic form, it was an open cockpit biplane seating two in tandem and it was powered with the Wright-Hisso (Hispano-Suiza) model E engine of 180 h.p.The cockpit coamings were heavily padded and the interior was quite bare except for necessities, the engine was completely uncowled and everything else was kept exposed as much as possible for ease of maintenance; to the last item it was designed to be functional. A type certificate number was issued to the Consolidated PT-1 in November of 1928, but this was merely a formality, in a sense, to acquire an A.T.C. "number". The A.T.C. numbers were much sought after because they were a badge of honor for an airplane, and undisputable proof of it's approved airworthiness.

In tracing it's development, we find that the PT-1 was a development from the earlier TW-3 that was first built by the Dayton-Wright Co. and then also built in an improved version by Consolidated Aircraft for an order of 20 ships. The PT-1 was extensively modified over the TW-3 design, and the many changes must brand it as a different airplane type entirely. Consolidated Aircraft was formed by Rueben Fleet who was Pres. and Col. Virginius E. Clark (who had been chief engineer at Dayton-Wright) was chief engineer for Consolidated. In 1923 at Buffalo, N.Y. "Consolidated" was formed as the successor to the liquidated Dayton-Wright Airplane Co. of Dayton, Ohio. Rueben Fleet, a shrewd man, surrounded himself with good able men and by 1925 they had received an order to build 100 Army trainers of the PT-1 type, one of the largest orders to be let at this time. A previous order for 50 of the PT-1 type was filled in units of 10 each. Sometime later the

Fig. 247. The Consolidated PT-1 with "Hisso" engine was no raving beauty, but it was a lovable airplane and trained many hundreds of Air Corps cadets.

Navy "air" placed a large order for the NY-1 type, (see discussion for A.T.C. #80 in this volume) and "Consolidated's" future was being well established. As everyone undoubtedly knows, it is now one of the largest airplane manufacturers in this country. All together, more than 220 of the PT-1 type were built, and then the Air Corps diverted most of them to service in the various National Guard units and standardized on the new Consolidated PT-3; see discussion for A.T.C. #83 in this volume.

Listed below are specifications and performance data for the "Hisso" powered Consolidated PT-1; span upper and lower 34'9", chord both 56", wing area 299 sq. ft., length 27'8", height 9'10", empty wt. 1805, useful load 713, crew wt. 434, gross wt. 2518 lb., max. speed 92, cruise 79, land 47, climb 690, ceiling 14,000 ft., gas cap. 40 gal., oil 5 gal., range 350 miles. The fuselage framework was built up of welded steel tubing, lightly faired to shape and fabric covered. The wings were built up of solid spruce spars and wood built-up ribs, also fabric covered. There were two fuel tanks, both in the center-section panel of the upper wing. The landing gear was of a rugged split-axle type and was built up of round, unfaired, steel tubing; as were the interplane bracing and center-section struts on most of the units. The interplane bracing wires were of standard braided aircraft cable, and there was an aileron on each panel connected together by a push-pull tube. The rudder and elevators were of the "balanced-horn" type. The landing gear used "DH-4" wheels and two spools of wound rubber shock-cord was used to take some of the jar out of the bumps.

Fig. 248. Early PT-1 of 1925, note "Jenny" type nose radiator and wing-tip skids.

There was an XPT-2, which was a PT-1 type that was powered with a "Whirlwind" J5 engine; this ship was then equipped with "Clark Y" wing panels and it became the XPT-3. When the Army Air Corps began taking delivery on the PT-3 type in 1928, all of the existing PT-1 type were being transferred to duty in the various National Guard Units.

A.T.C. #80
(11-28)
CONSOLIDATED "HUSKY", NY-1

Fig. 249. The rugged NY-1 was built to stand the abuse suffered in pilot training.

The Consolidated NY-1 development was basically a type typical to the PT-1, except that it was modified into a "Navy type" primary trainer. Proposed in the Fall of 1925, the Consolidated design underwent only slight modification to adapt it to Navy training service, the main change was the installation of a new production air-cooled radial engine in place of the war-surplus "vee-type" water-cooled engine used by the Army Air Corps. Apparently the requisites for a Navy type trainer were somewhat different. The Navy "air" was a staunch supporter of the air-cooled "radial" type engine and it's belief and great trust in this particular type was later justified beyond question. Con-

Fig. 250. A J5 powered NY-1 in service at Pensacola air station, where the Navy's fledglings were trained.

Fig. 251. The rugged and simple character of the NY-1 is clearly reflected here.

solidated's initial order for the NY-1 was for 40 ships early in 1926.

As pictured here in the various views, the NY-1 "Husky" was a 2 place open cockpit biplane that seated two in tandem, and it was powered with a 9 cylinder Wright "Whirlwind" engine; first with the J4 series of 200 h.p., and later with the J5 series of 220 h.p. Other than the change in engines, the NY-1 also had a redesigned tail-group of a bit more area, and was used extensively as a seaplane trainer. This transition was accomplished by the removal of the wheeled landing gear and the installation of a large central "float" in it's place. There was a small wing-tip float on each side for lateral balance. The NY-1 was also very rugged in character and was partly in keeping with a comment someone once stated that a training ship is an airplane who's main requisite was the ability to stand up under the severe, even though unintentional, abuse dealt it by wide-eyed fledgling pilots in trying to learn to fly! The "Consolidated" trainers were designed and built with this in mind.

The type certificate number for the Consolidated NY-1 was issued in November of 1928, by this time it was already being replaced in service by the improved NY-2; see discussion for A.T.C. #81 in this volume. The basic "Husky" design later became a pattern for the development of the "Husky Jr.", a civilian-type training airplane that led to the fabulous "Fleet" biplane.

Listed below are specifications and performance data for the "Whirlwind" powered Consolidated NY-1; span upper and lower 34'6", chord both 56", wing area 297 sq.ft., length 27'10", height 9'10", empty wt. 1773, useful load 722, crew weight 450, gross wt. 2495 lb., max. speed 105, cruise 85, land 46, climb 765, ceiling 15,000 ft., gas cap. 40 gal., oil 3 gal., range 300 miles. For construction details of the fuselage framework, wing framework, tail-group, etc., see previous discussion of the PT-1. The only construction details that would differ would be the engine mount, and the addition of "float" fittings. A metal propeller was sometimes used, and a hand-crank inertia type engine starter was usually installed for seaplane work. The performance of the seaplane version was perhaps as much as 8% less than the landplane version.

Fig. 252. The Consolidated NY-1 as a seaplane trainer.

A.T.C. #81
(11-28)
CONSOLIDATED "HUSKY", NY-2

Fig. 253. The Consolidated NY-2, powered with Wright J5 engine.

The Consolidated model NY-2 was basically a modification of the earlier NY-1 type and was developed to be used as a primary trainer both for land and sea, and was also used as a secondary trainer in teaching embryo pilots the intricacies of aerial observation work and aerial gunnery. Mountings in the forward portion were provided for fixed machine-guns to be used in pilot training, and swivel mounted flexible machine-guns were mounted in the rear cockpit for training of the observer-gunner. The NY-2 was used extensively for seaplane training and the wheeled landing gear was quickly demountable to take a large central float, with the addition of a small wing-tip float outboard on each wing. Typical of the earlier NY-1 to a great extent, the NY-2 was also a two place open cockpit biplane seating two in tandem but it was a slightly larger airplane; it had added wing area to compensate for the added weight of the guns and other extra equipment used in secondary phases of pilot training. The powerplant for the model NY-2 was usually the 9 cylinder Wright "Whirlwind" J5 engine of 220 h.p., the increased weight of this model caused a higher power loading and consequently performance suffered in some respects. The NY-2 flew and handled well, and stood up admirably under the severe abuse and misuse often subjected to by fledgling pilots.

The type certificate number for the Consolidated NY-2 was issued in November of 1928, and in one or two cases was also built in a modified "commercial" version. It was a ship of this type, loaned by the N.A.C.A. for this venture, that was used by "Jimmie" Doolittle in Sept. of 1929 at Mitchel Field on his "Blind Flight" over a 15 mile triangular course that included a take-off and landing! This sort of thing we might judge as commonplace today, but it was quite a remarkable feat back in the year of 1929.

Listed below are specifications and performance data for the "Whirlwind" J5 powered Consolidated NY-2; span upper and lower 40', chord both 60", wing area 370 sq. ft., airfoil Clark Y, length 27'10", height 9'11", (as landplane); empty wt. 1970, useful load 700, crew wt. 460, gross wt. 2670 lb., max. speed 95, cruise 80, land 48, climb 690, ceiling 12,000 ft., gas cap. 40 gal., oil 3

Fig. 254. The NY-2 was used extensively for seaplane training, shown here with "Edo" metal main float.

gal., range 250 miles. The following figures are for seaplane; empty wt. 2145, useful load 698, crew wt. 458, gross wt. 2843 lb., max. speed 90, cruise 75, land 52, climb 630, ceiling 11,000 ft., range 210 miles. The fuselage framework was built up of welded carbon-steel tubing in three sections, the rear section was braced with steel tie-rods and the two forward sections were braced with steel tubing, the fuselage was then lightly faired to shape and fabric covered. The wing framework was built up of solid spruce spars and wood built-up ribs, also fabric covered. The wing panels were now of heavy "Clark Y" section and the ailerons were mounted on offset hinges to increase aileron effectiveness and improve lateral control. The fuel supply of 40 gal. was in a tank mounted in the center-section panel of the upper wing. The rugged landing gear was built up of round unfaired steel tubing and used "DH-4" wheels. The wing "N" struts were of chrome-moly steel tubing while the center-section struts were of carbon-steel tubing; all interplane bracing wires were of standard round braided aircraft cable, all movable surfaces were cable operated. Both rudder and elevators were of the "balanced horn" type. The fabric covered tail-group was built up of welded steel tubing, the fin was ground adjustable and the horizontal stabilizer was adjustable in flight. The modified "commercial version" of the NY-2 incorporated a few niceties in it's make up such as, streamlined struts for the interplane bracing, the landing gear vees were faired, the fuselage was faired to a fuller shape, and the cockpit coamings were minus the oversize crash-pads; all this provided a noticeable increase in performance. Wheel brakes, metal propeller, and an inertia-type engine starter were available.

Fig. 255. View taken Aug. 1931 of NY-2 at Pensacola Air Station. Some of this type served unto the late "thirties".

A.T.C. #82
(11-28)
CONSOLIDATED "COURIER", 0-17

Fig. 256. The Consolidated "Courier" model 0-17, powered with Wright "Whirlwind" J5 engine.

The Consolidated model 0-17 was basically a slightly modified PT-3 or NY-2 type that was especially redesigned and built for use in the National Guard units as an observation and ground liason trainer, or used for gunnery, radio contact, and photo work in cooperation with infantry units. The 0-17 was the lesser known of this Consolidated series of "service trainers", but actually performed many more diversified services. In it's basic form, it was also a two place open cockpit biplane seating two in tandem, and was powered with a 9 cylinder Wright "Whirlwind" J5 engine of 220 h.p. The model 0-17 was called the "Courier", and differed mostly in that it was "cleaned up" quite a bit over it's counterparts. The fuselage was faired out much better, the cockpits were modified, in some cases eliminating the excessive padding of the cockpit coaming that was so typical of the PT and NY series. All interplane bracing, wires and struts, were of streamlined section; the nose-cowling was faired into a propeller "spinner".

The early "Courier" of 1927 was no doubt modified from a PT-1, because it still had the PT-1 tail-group; later "Couriers" had PT-3 type tail-groups and some even had the NY-2 type. These were either modifications that took place during the evolution of the type, or else "the boys" just hung on what they grabbed first. A type certificate number was issued to the Consolidated 0-17 in November of 1928 and about 30 of this model were built. In view of the fact that not very many of the "Couriers" were built, they were quite rare and spread pretty thinly about the country. A number of the early "Army and Navy" types were given A.T.C. numbers by request of the manufacturer, but this practice was later deemed unnecessary and was discontinued.

Listed below are specifications and performance data for the "Whirlwind" J5 powered Consolidated "Courier" model 0-17; span upper and lower 34'6", chord both 56", wing area 297 sq.ft., airfoil Clark Y, length 27'11", height 9'9", empty wt. 1881, useful load 842, crew wt. 480, gross wt. 2723 lb., max. speed 118, cruise 100, land 48, climb 865, ceiling 12,000 ft., gas cap. 40-60 gal., oil 3 gal., range 350-550 miles. The construction details of the 0-17 were more or less similiar to the other Consolidated "service trainers", refer to previous discussions. The landing gear on the 0-17 was modified somewhat, and the tail-skid was steerable in connection with the rudder pedals. A metal propeller and a hand-crank inertia type engine starter were usually provided. In case some are not familiar with the inertia-type engine starter, it was just a high geared fly-wheel that was brought up to an extremely high r.p.m., either by hand-crank or electrically, and then this flywheel energy was transferred to the engine shaft to turn it over for firing. The high "whine" of the fly-wheel when brought up to speed was always intriguing. For the next development in the "Consolidated" series, see discussion for A.T.C. #83 in this volume.

Fig. 257. Excellent view of the "Courier" showing fuel tanks in upper wing, aileron control detail, and note "scarf ring" on rear cockpit for machine-guns.

A.T.C. #83
(11-28)
CONSOLIDATED, PT-3 & PT-3A

Fig. 258. The Consolidated PT-3 with Wright J5 engine was the standard primary trainer in the Air Corps for many years.

The Consolidated PT-3 series as used by the Air Corps were basically a modified and modernized version of the earlier PT-1, the main change was the installation of a 9 cylinder Wright "Whirlwind" engine of the J5 series that was rated at 220 h.p. at 1800 r.p.m. In it's basic form, the configuration of the PT-3 was almost identical to the PT-1 except for the shape of the tail-group, "Clark Y" wing panels, and a few other minor changes such as a light fairing out of the fuselage and a cowling around the engine section. Up to now the Army "Air Service" and "Air Corps" had been more or less honor-bound to use liquid-cooled engines in all of their airplane types, but was beginning to relent somewhat and take a flyer in equipping some of their airplanes with air-cooled engines.

The model PT-3 was also an open cockpit biplane seating two in tandem, and it was just as amiable as the beloved PT-1, if not more so, and was also very rugged in character. These "Consolidated" trainers were a lean and sinewy breed that performed their intended duties in a fine manner and could absorb a terrific amount of abuse without much complaint. Although the PT-3 series didn't appear quite as stark as the PT-1, everything was kept quite simple and still handy for efficient servicing and maintenance; an attribute quite necessary in flight training work to keep the "hangar time" of an airplane at a minimum. With it's advent into full-scale service, first used at Brooks Field in Texas, the PT-3 became the standard primary trainer for the Air Corps and the remaining PT-1's were retired from use and transferred to service in various National Guard units. Even though operating under exacting conditions, the PT-3 had proved itself well and gave many years of faithful service. It also went into a modification known as the PT-3A, which only had some minor changes, and altogether some 250 of these airplanes were built. The PT-3 was the first "Corps" airplane to experiment with blind-flying "hoods" over the student's cockpit for training to fly entirely by instruments, several different versions of these "hoods" were tried. A type certificate number was issued to the PT-3 and PT-3A in November of 1928 and they were manufactured by the Consolidated Aircraft

Corp. at Buffalo, New York.

Listed below are specifications and performance data for the "Whirlwind" J5 powered Consolidated PT-3 and PT-3A; span upper and lower 34'6", chord both 56", wing area 297 sq. ft., airfoil Clark Y, length 25'9", height 9'9", empty wt. 1747, useful load 733, crew wt. 480, gross wt. 2480 lb., max speed 105, cruise 90, land 47, climb 765, ceiling 14,900 ft., gas cap. 40 gal., oil 3 gal., range 310 miles. For construction details of the fuselage framework and wing panels, see "Consolidated" types in previous discussions. The fabric covered tail-group was built up of welded steel tubing, the fin was ground adjustable and the horizontal stabilizer was adjustable in flight. The rudder used an "overhanging horn" for aerodynamic balance but the elevators did not. All struts, wires, landing gear, etc., were comparable to the other "Consolidated" trainer series. The fuel supply was held in two tanks of 20 gal. each, these were mounted in the center-section panel of the upper wing. The early color scheme for the PT-3 was a white fuselage with orange-yellow wings and tail-group, later this was changed to an olive-drab fuselage, struts, and landing gear; with wings and tail surfaces of Army yellow.

A word or two pertaining to the lineage of these "Consolidated" PT types; the PT-1 was developed from the TW-3, then a "Whirlwind" J5 was installed in a PT-1 for test and it became the XPT-2, new wing panels of "Clark Y" section were then installed on this version and it became the XPT-3, numerous modifications were added to the production model and it became a PT-3; still further modification made it into a PT-3A, which was the last of the series except for those test versions just mentioned. A PT-3 was to be tested with a 4 cylinder "Caminez" engine as an XPT-4 but was cancelled. In 1929, a 6 cylinder Curtiss "Challenger" engine of 170 h.p. was installed in an XPT-3 and it became the XPT-5; in 1930, a PT-3A was converted to take the 9 cylinder Packard "Diesel" engine of 225 h.p. and it became the XPT-8A. There were other "Consolidated" trainer types after this, but they finally gave way to the "Stearman" series, notably the PT-13 which then became the standard "service trainer" for the Air Corps. The next "Consolidated" development to come along was the nimble "Husky Jr.", see discussion for A.T.C. #84 in this volume.

Fig. 259. Consolidated PT-3 with Wright J5 engine.

A.T.C. #84
(11-28)
CONSOLIDATED "HUSKY JR."

Fig. 260. The "Husky Jr." was a sport-trainer that became the basis for the fabulous "Fleet".

The Consolidated Model 14 or "Husky Jr.", as it was better known, was a natural development from the Consolidated PT and NY series, and clearly showed it's ancestral ties in many ways; retaining all of the rugged lines and characteristic features. This model was an experiment and was more or less a probe into the civilian market to see if this type of sport-trainer airplane would sell; the "Husky Jr." was some time in catching on but it did lay a good groundwork for it's future counterpart, the fabulous "Fleet" sport-trainer.

As pictured here in the various views, it was a 2 place open cockpit biplane that seated two in tandem in one elongated cockpit section, with only a windshield frame separating the two occupants. As on the PT and NY, everything vain such as finery, was kept to a minimum; mainly stressing efficient simplicity. The powerplant for the Model 14 was the 7 cylinder Warner "Scarab" engine of 110 h.p., which gave it a good reserve of power for an effortless and lively performance. The sprightly nature of this little airplane naturally called for a rugged construction and the "Husky Jr.", as the name implies, met this qualification admirably with flying colors; an attribute that was later so ably demonstrated and confirmed by the "shenanigans" of the "Fleet". By this we mean the many "outside loop" records and such, and a reputation in the industry and flying-circles alike as practically an "unbreakable" airplane!

The type certificate number for the Model 14 "Husky Jr." was issued in November of 1928 and it's manufacture was continued into 1929. With a modification of the elongated open cockpit into 2 individual open cockpits and a few other minor changes, this airplane later became the "Fleet" Model 1. For details of this type, see discussion for A.T.C. #122. The "Husky Jr." was manufactured at Buffalo, New York, by the Consolidated Aircraft Corp., Rueben Fleet was Pres. and Gen. Mgr. and Lawrence "Larry" Bell, later of "Bell Aircraft" fame, was the General Sales Mgr.

Listed below are specifications and performance data for the Warner "Scarab" powered Consolidated "Husky Jr."; span upper and lower 28', chord both 45", wing area 195 sq. ft., airfoil Clark Y, length 20'9", height 7'10", empty wt. 967, useful load 474, payload 165, gross wt. 1450 lb., max. speed 105 cruise 90, land 40, climb 850, ceiling 13,500 ft., gas cap. 24 gal., oil 2.5 gal., range 350 miles. A short time later the "Husky Jr.'s" certificate was ammended to allow a useful load of 554 lb. and a gross load of 1530 lb., this allowed an additional payload of 80 lb. for a total of 245 lb. instead of 165 lb. This weight increase had very little effect on the overall performance. The fuselage framework was built up of welded chrome-moly steel tubing, lightly faired to shape and fabric covered. The extra rugged wing framework was built up of heavy solid spruce spars and stamped-out aluminum alloy ribs, also fabric covered. All wing and center-section struts were of heavy gauge streamlined steel tubing, with interplane bracing wires of heavy steel in streamlined section. The "Husky Jr." introduced the cross-axle type "oleo" landing gear that was retained in the various "Fleet" models. The upper wing was built in a continuous section and the fuel tank was mounted in the center portion. The tail-group was built up of welded steel tubing and sheet steel ribs, fabric covered. All tail surfaces were of heavy cross-section, and the horizontal stabilizer was of a "lifting section", a feature that characterized the "Fleet" models throughout the whole series. The fin was ground adjustable and the horizontal stabilizer was adjustable in flight. Ailerons were on the lower panels only and were operated through a positive acting torque tube and bellcrank action. The "Husky Jr." was light and quick on the controls, capable of the whole retinue of aerobatic manuvers that were only limited by the pilot's ability and fortitude.

Fig. 261. "Husky Jr." with Warner engine, note dihedral angle and cross-axle landing gear.

A.T.C. #85
(11-28)
HAMILTON "METALPLANE", H-45

Fig. 262. Hamilton "Metalplane" model H-45 with Wasp engine. Note covering of corrugated "Alclad" metal skin.

The sturdy looking Hamilton "Metalplane", model H-45, was a high wing all-metal cabin monoplane of simple and efficient design that seated 8 places, and was powered with a 9 cylinder Pratt & Whitney "Wasp" engine of 410-450 h.p. As pictured here in the various views, we see it's thick full-cantilever wing, it's wide tread long-leg landing gear, and it's corrugated metal skin that was used as a covering throughout; all these were features designed to stand up under continuous service with a minimum amount of maintenance. Designed primarily to transport passengers and cargo, this model had enjoyed enthusiastic acceptance on a few of the early air-lines. Two H-45 "Metalplanes" were added to the growing "Northwest Airways" fleet in Sept. of 1928 to be used on their St. Paul to Chicago route. Later, additional "Hamiltons" were added and service was extended progressively into the Dakotas, into Montana, and finally clear into Seattle. By 1934, "Northwest" had 8 "Hamiltons" serving in their fleet. The "Hamilton"metal monoplane is probably more often identified with "NWA" (Northwest Airways), but they were also in use by Universal Air Lines, Isthmian Airways, and some others.

The Hamilton "Metalplane" actually dates back to 1926 and was built by the Hamilton Aero Mfg. Co., a firm that was then already quite prominent and famous for it's "Hamilton" wood propellers, and metal pontoons and flying boat hulls that were built of the new alloy called "dural". Their original "Metalplane" series were designed by James S. McDonnell, who in more recent years became best known for his "Phantom", "Banshee", and "Voodoo" jet powered fighter aircraft. In it's basic form, the original "Hamilton" type was an all-metal "shoulder wing" cabin monoplane with an internally braced full cantilever wing and was of the type as shown in one of the views. The pilot sat in an open cockpit just ahead of the wing leading edge and the cabin section under the wing seated 4 passengers. It was first powered with a Wright "Whirlwind" J4 engine of 220 h.p., which was later replaced with a "Whirlwind" of the J5 series. This type was later modified still further with the installation of a P & W "Wasp" engine of 400 h.p. In 1927, a J5 powered "Metalplane, the only one built in that year, was flown by Randolph Page and took a second place to "Eddie" Stinson in the Ford Air Tour of that year. Later that year in 1927, the same plane flown by John Miller won highest honors in the transport division of the Air Races held at Spokane by winning first place in both efficiency contests.

Fig. 263. Northwest Airways was the first to use the Hamilton "Metalplane". Here is the "Wasp" powered H-45, Chas. "Speed" Holman their operations manager is on right.

Although this early design was sound and efficient, it tended to be somewhat complicated so this configuration was discarded in favor of the normal high wing type, which was more practical. The new high wing versions were designed by Prof. John Ackerman and they were the first of the all-metal airplanes to be certificated, but not the first to be built here in the U.S.A. This honor would no doubt go to the Stout "Pullman" transport of 1924, which eventually evolved into the famous "Ford Tri-Motor". All of the Hamilton "Metalplanes", since the very first, were of all-metal construction throughout, using an aluminum alloy framework and all assemblies were covered with a corrugated "dural" metal skin. This method of construction was also typical of the "Stout" and "Ford" designs, and later on such other airplanes as the "Prudden", "Thaden", and others.

The type certificate number for the Hamilton "Metalplane" model H-45 was issued in November of 1928, and re-issued for added wing area and some other slight modifications in May of 1929. An earlier model, the H-43, was built under a Grp. 2 approval that was

Fig. 264. Passengers boarding a Northwest Airways "Hamilton" H-45, Chas. "Speed" Holman on right.

Fig. 265. 1927 version of the Hamilton "Metalplane" powered with Wright J5 engine. This ship won 2nd place in 1927 Ford Air Tour, "Tom" Hamilton on right.

issued in August of 1928. The H-43 and the H-45 were manufactured by the Hamilton Metalplane Co. at Milwaukee, Wis., Thomas F. "Tom" Hamilton was the Pres. "Hamilton" became a division of the Boeing Airplane Co. in the latter part of 1928. The next development of the "Metalplane" series was the improved model H-47, see discussion for A.T.C. #94 in this volume.

Listed below are specifications and performance data for the "Wasp" powered Hamilton "Metalplane" model H-45; wing span 51'1", chord at root 108", chord at tip 75", wing area 360 sq. ft., airfoil Clark Y Mod., length 34'10", height 8'7", empty wt. 2900, useful load 2500, payload 1110, gross wt. 5400 lb., max speed 140, cruise 115, land 50, climb 900, ceiling 13,500 ft. The following figures are for the later model; wing span 54'5", chord at root and tip was the same, wing area 387 sq. ft., length 34'8", height 9'3", empty wt. 3342, useful load 2408, payload 1368, gross wt. 5750 lb., max. speed 138 cruise 115, land 50, climb 850, ceiling 13,000 ft., gas cap. 140 gal., oil 10 gal., range 675 miles. The following figures were allowed on the ammended certificate; empty wt. 3639, useful load 2111, payload 1000, gross wt. 5750 lb., powerplant was now the "Wasp" of 450 h.p., there was no noticeable difference in the performance figures. Price at the factory was $23,200 and later raised to $24,500. The fuselage framework was built up of open section riveted Alclad, and covered with a corrugated skin also of Alclad. The wing framework was built up of truss-type spars of riveted "dural" tubing and the ribs were of stamped-out dural, the completed framework was covered with corrugated dural sheet. The tail-group was built up in like manner, the fin was ground adjustable and the horizontal stabilizer was adjustable in flight. Wheel brakes, metal propeller, and inertia-type engine starter were standard equipment, the wings were wired for lights. The model H-45 was also available as a seaplane equipped with "Hamilton" metal floats.

A.T.C. #86
(11-28)
LAIRD "COMMERCIAL", LC-B

Fig. 266. Leveled at the sportsman-pilot, the Laird LC-B was ideal for cross-country jaunts.

The "Laird Commercial", or any "Laird" for that matter, possessed a magnetic personality of subtle flavor, and was the type of airplane that would cause most anybody to want to walk over and take a real good look of careful scrutiny; they had that sort of compelling charm and quiet beauty that was hard to pass up. Then too, the type was not too plentiful in some areas, so it was somewhat of a rare treat to look one over up close. Let's say, not particularly a rare type, but they were most always in limited production because they were more or less a custom built airplane. The "Laird" was usually built to suit the customer's particular fancies and requirements; likened to a fine piece of precision machinery, they were built unhurried and with the greatest of care. E. M. Laird proudly advertised his airplanes as the "Thoroughbreds of the Air" and this could hardly be disputed, because they had proven time and time again that they had every justification to make such a statement.

The first thing that impressed one about the "Laird" was it's unusual color scheme and very handsome finish, which was a deep, rich black on the fuselage and it's lustrous gold colored wings of the richest looking hand-rubbed finish you ever saw! The model "LC-B" was not otherwise too unusual; as pictured here it was a typical 3 place open cockpit biplane of conventional lay-out that was proportioned in good harmony, both in appearance and aerodynamically. The standard powerplant installation was the 9 cylinder Wright "Whirlwind" J5 engine of 220 h.p. Performance in this combination was just slightly better than the average; the "Laird Commercial" were extremely sure-footed and delightfully responsive, they were a real joy to fly.

"Laird" history goes way back, at least back to the turn of the "twenties" when they were building the 3 place OX-5 powered "Laird-Swallow", an airplane that in description could be called a sort of "cleaned up Jenny". It was hailed as America's first "commercial airplane" and in turn became the ancestral background of the later day "Swallow", or "New Swallow", as it was first called. E.M. Laird, somewhat disgruntled by events, had left the organization by now, and the "New Swallow" had been designed by Lloyd Stearman. Right on the heels of the "Swallow", the first "Laird Commercial" appeared later in 1924. As shown here, the

Fig. 267. Early example of the "Laird Commercial", powered with Curtiss C-6 engine. This ship flown by Hy Etten, was in Ford Air Tour of 1925.

first of these types were fairly typical to the LC-B, except that they had an axle-type landing gear and water-cooled engines. Some were powered with the Curtiss OX-5 engine of 90 h.p., and some were powered with the 6 cylinder in-line Curtiss C-6 engine of 160 h.p. One of these early "Lairds", shown here, was flown in the first Ford Air Tour of 1925 by Hy Etten and did very well in a show of performance.

The "Whirlwind J4" powered Laird-Commercial was out early in 1926 and was one of the earliest pioneer carriers on the contract airmail routes. Charles Dickinson, a lovable old man with plenty of money, formed a line that carried the air-mail on C.A.M. #9 in J4 powered Laird-Commercials on the Chicago, St. Paul, and Minneapolis route. This early ventured operated only a few months, went broke, and was then taken over by "Northwest Airways" in Nov. of 1926. Two of Chas. Dickinson's original "Laird-Commercials" were later modified to some extent and were entered in the cross-country Air Derby of 1927 from New York to Spokane. They were flown by Chas. "Speed" Holman and E. E. Ballough to a one-two finish in a hotly contested race. Holman's plane (C-240), and that of E. E. Ballough (C-110), are both shown here. Two "Laird-Commercials" of an improved type were purchased by the "Aero. Div." of the Dept. of Commerce later in 1927 for use by their inspectors in field work, this selection by the "Department" was also good testimony of the Laird reputation for excellence.

The type certificate number for the model LC-B was issued in Nov. of 1928. A Grp. 2 approval numbered 2-17 was issued in August of 1928 for all of the existing "Laird-Commercials" that were manufactured prior to October of 1928; all subsequent airplanes of this model, after this date, were built under the A.T.C. #86.

E. M. Laird, a pilot, engineer, and designer of the old school; designed and developed, and most often flight-tested the various models that were built throughout the years. E. M. Laird, who's full name hardly ever known was Emil Matthew Laird, but he was much better known as just plain "Matty" Laird; the inspiring president of the E. M. Laird Airplane Co. of Chicago, Ill. "Matty" had a brother named Charles who was building the "Whippoorwill" in Wichita, Kansas in 1928, but this venture was of but short duration. "Matty" Laird never got rich selling airplanes, but he sure had fun designing and building them, he loved what he was doing. To him, and to many others we could mention, aviation

Fig. 268. J4 powered "Laird Commercial", flown by E. E. Ballaugh (right background) to second place in 1927 Air Derby from N. Y. to Spokane.

was a way of life more than a business.

Listed below are specifications and performance data for the "Whirlwind" J5 powered "Laird Commercial" model LC-B; span upper 34', span lower 30'6", chord upper 65", chord lower 54", wing area 295 sq. ft., airfoil "Laird #1" (some later versions had "Laird #2), length 23'9", height 9'3", empty wt. 1800, useful load 1050, payload 420, gross wt. 2850 lb., max. speed 130, cruise 110, land 48, climb 1000, ceiling 19,000 ft., gas cap. 76 gal., oil 7 gal, range 650+ miles. Price at the factory was $9850. The fuselage framework was built up of "dural" tubing held together by steel-clamped joints and trussed with steel tie-rods, the framework was amply faired and fabric covered. The wing panels were built up of solid spruce spars and spruce and plywood built-up ribs, also fabric covered. The fuel supply was carried in 2 tanks, the main tank was in the fuselage and the other tank was mounted in the center-section of the upper wing. The landing gear was of the normal split-axle type and was fitted with individual wheel brakes. The fabric covered tail-group was built up of welded steel tubing, with exception of the horizontal stabilizer which was built up of wood spars and ribs. The fin was ground adjustable and the horizontal stabilizer was adjustable in flight. An inertia-type hand crank engine starter and a metal propeller were standard equipment. Leveled at the sportsman-pilot, the LC-B was more than ideal for extensive cross-country jaunts; with provisions for carrying some 90 pounds of baggage and a cruising range of well over 6 hours. The next "Laird" development was the J4 powered Laird "Speed-Wing" model LC-RJ200, see discussion for A.T.C. #152.

Fig. 269. This famous J4 powered "Laird-Commercial" was flown by Chas. "Speed" Holman to first place in the 1927 Air Derby from N. Y. to Spokane, Wash.

A.T.C. #87
(11-28)
FORD "TRI-MOTOR", 4-AT-B

Fig. 270. 1928 Ford "Tri-Motor" in the early 4-AT-B version.

The time-worn and lovable old Ford "Tri-Motor" is probably now one of the most well-known and well-remembered airplanes of all times. As many may already know or will recall, the "Ford" came to be fondly referred to as the "Tin Goose"; mainly because of it's ample proportions done up in an all-metal construction. It's corrugated metal skin covering was a distinguishing feature that set it well apart from the general run of airplanes as they were built back in this day.

Going back for a moment into early "Ford" airplane history we might truthfully say that the Ford "Tri-Motor" design was actually prompted into being by the genial "Tony" Fokker and his own "Fokker Tri-Motor", a plane that was brought out and demonstrated nation-wide in 1925. Henry Ford was greatly impressed by the success of Fokker's airplane and it's many capabilities. Deciding in good judgement that the future of aviation must lie in "transport", he suggested to his engineering staff that they should enter the field and also design and develop a "tri-motor", but they must definitely go Fokker one better! As it turned out though, their first attempt under the added strain of compulsion was fartoo radical and did not produce the results so eagerly expected. Their second try, the 4-AT, of a more conventional configuration was much better, and became the prototype for the subsequent models of which quite a good number were built.

The first Ford "Tri-Motor", as shown here, was actually a development from the Stout "Pullman", sometimes called the "Airsedan"; a transport type which came out late in 1924 and at the time of the proposed development of a "tri-motor" type, was already being built in the Ford shops. The "Pullman" type is also shown here in one of the views. The "Stout" company after a time of benevolent support by Ford, was absorbed into the Ford organization in August of 1925 and immediately resumed an accelerated production on the "Liberty 12" powered "Pullman". The Stout "Pullman", for all intents, was considered as the first all-metal commercial airplane to be built in the U.S.A. In it's basic form, it was a single engined high wing cabin monoplane that had a full cantilever wing and it was powered with the venerable 12 cylinder "Liberty" engine of 400 h.p. It's framework was of alumimum alloy in an all-metal construction and it used a corrugated metal skin covering that was to become so characteristic of the "Tin Goose".

Having such a good basis to start from, Wm. B. Stout and his staff of engineers which included men like Geo. Prudden, Tom Towle, Otto Koppen, and others; all men of some degree of genius, tackled the problem with

Fig. 271. Ford's first "tri-motor" (3-AT) was an unusual configuration that was developed from the basic "Pullman" design. Only one built.

relish. Together they redesigned the "Pullman" type considerably to take the power of three Wright "Whirlwind" J4 engines of 200 h.p. each; and so it was that the Ford-Stout "Tri-Motor" was born. This airplane, in a sense, was of a very unusual configuration, it is shown here for comparison; after a few test flights it was decided with some misgivings that it would not come up to expectations and requirements, so a new design was announced in December of 1925. The first actual "Tri-Motor" as we came to know it was the "4-AT", which was formally introduced sometime in June of 1926, with flight tests completed a month or so later. The passenger's cabin on the 4-AT was fully enclosed to carry 8 people comfortably but the pilot's cockpit, by preference, was still open and was placed just ahead of the wing's leading edge as on the "Pullman".

In 1925 a Ford-Stout "Pullman" was flown in the Ford Air Tour by E. G. "Eddie" Hamilton, and later in 1926 the first 4-AT tri-motor was flown by Leroy Manning as a participant in the Ford Air Tour of that year; both airplanes did well although the 4-AT failed to complete the circuit. In 1927 the Ford "Tri-Motor" had been remodeled somewhat and some were later powered with 3 of the new "Whirlwind" J5 engines that were built in that year. Dean Mulford flew one of these later type in the 1927 Ford Air Tour and finished in a 4th place. The model 4-AT-A, built throughout 1927, had a seating arrangement for 12 to 14 places and was powered with the Wright J4 engines first and later on was powered with 3 of the new J5 engines; some 14 of this model were built, and licensed

Fig. 272. This was the Stout (2-AT) "Pullman" powered with 12 cyl. "Liberty" engine. Put into service on Florida Airways in 1925.

Fig. 273. A later version of the 4-AT-B.

later under a Grp. 2 approval numbered 2-9 which was issued in Sept. of 1928. Ford launched serious production of his "Tri-Motors" on about October of 1928 and by the end of a year's time had manufactured some 135 airplanes in the various series and modifications.

The Ford "Tri-Motor" model 4-AT-B, as featured in this discussion was first issued a Grp. 2 approval numbered 2-10 but this was superceded by A.T.C. #87, which was issued in November of 1928. Typical of the earlier models to a great degree, the 4-AT-B was also an all-metal high wing cabin monoplane with a full cantilever wing that was now of some greater span and more area. It was powered with 3 Wright J5 engines of 220 h.p. each; one engine was mounted in the nose of the fuselage and the other two engines were mounted in strut braced nacelles, one under each wing. It was only slightly modified over preceding models and also had seating for 12 to 14 places; usually 12 passengers and a crew of two. The 4-AT-B was soon placed in regular service on numerous air-lines and also was quite popular with big business; the 4-AT-B is shown here in various views. This model also became the "C-3" in the Army Air Corps and was used as a utility-cargo transport, one was procured in 1928. A Ford "Tri-Motor", the last of the 4-AT-A series, was flown by Frank Hawks (later of speed dash fame) in the 1928 National Air Tour (formerly the Ford Air Tour) and finished in 2nd place to a sprightly "Whirlwind-Waco" flown by John P. Wood. By all criteria this was a very creditable showing for a ship of this size and type.

The mellow old "Tin Goose" was blessed with a compatible nature and had a capable air about her that seemed to instill friendship and immediate trust in all who rode or flew

Fig. 274. This was Ford's second "Tri-motor" design (4-AT), note open cockpit for pilot.

Fig. 275. This was Wm. B. Stout's first airplane, the (1-AS) "Air Sedan". It had seating for 4 and was powered with OX-5 engine. Became basis for Liberty powered "Pullman".

her! This attribute has stood the "goose" in good stead throughout the years and has preserved her popularity even to this day.

Listed below are specifications and performance data for the "Whirlwind" J5 powered Ford "Tri-Motor" model 4-AT-B; wing span 74', chord at root 154", chord at tip 92", wing area 785 sq. ft., length 49'10", height 12'8", empty wt. 6169, useful load 3961, payload 2200, gross wt. 10,130 lb., max. speed 114, cruise 95, land 55, climb 750, ceiling 12,000 ft., gas cap. 235 gal., oil 15 gal., range 520 miles. The Ford "Tri-Motors" were manufactured by the Stout Metal Airplane Co., a division of the Ford Motor Co. at Dearborn, Mich. Wm. B. Mayo was Ford Motor's chief engineer with Wm. B. Stout as the chief engineer of the airplane division. Tom Towle (later of Towle Amphibian fame) was Stout's chief of engineering; Geo. Prudden had left in 1927 to develop his light "tri-motor" in San Diego, Calif.

The fuselage framework on the 4-AT-B was built up of "dural" channels and sections that were riveted together, then covered with an "Alclad" aluminum alloy skin that was corrugated for added strength and stiffness. The wing framework was entirely of aluminum alloy sections riveted together and also covered with a corrugated "Alclad" skin. The tail-group was built up in like manner. "Alclad" was a dural metal skin that was covered with a thin coating of pure aluminum to resist corrosion, the inner structure of the wing and fuselage was not built up of "Alclad" so they were protected from corrosion by a special applied preventative coating. The fuel supply was carried in tanks that were mounted in the wing, the oil tanks were in the engine nacelles. Inertia-type engine starters, wheel brakes, and metal propellers were standard equipment. The next development in the "Tri-Motor" 4-AT series was the "Whirlwind" J6-300 powered model 4-AT-E.

A.T.C. #88
(11-28)
KREIDER-REISNER "CHALLENGER", C-4A

Fig. 276. The "Comet" powered Kreider-Reisner "Challenger" biplane model C-4A (KR-34A).

This model in the Kreider-Reisner "Challenger" series, as pictured here, was also a 3 place open cockpit biplane of the light commercial type. A development that was basically similiar to the earlier model C-1 and C-2 (KR-31) that was powered with an OX-5 engine. That is to say, it was similiar except for the following changes; the landing gear on the C-4A had been modified to the split-axle type and used rubber "shock-rings" working in compression, to absorb landing loads. There were now two positive-actuated ailerons on the lower wing panels only, and it was powered with the very interesting 7 cylinder "Comet" air-cooled "radial" engine of 115-130 h.p. The first few models of the improved C-4A type to come off the line were powered with the newly introduced "Comet 115", but the "Comet" engine soon received a rating boost to 130 h.p. at 1825 r.p.m. The "Challenger" biplanes were of a fairly average performance, as would be expected from a plane of this type, but were possessed of good habits and flew exceptionally well.

An earlier experimental version of the "Challenger" biplane was powered with the 4 cylinder "Caminez" engine of 120-135 h.p. This plane shown here, was flown by Amos Kreider in the 1928 National Air Tour and finished in 26th place; a shake-down cruise that brought forth many of the problems still inherent in this engine design, and more or less marked the early departure of this engine from the scene. The type certificate number for the model C-4A (KR-34A) was issued in Nov. of 1928 and it was manufactured by the Kreider-Reisner Aircraft Co. at Hagerstown, Maryland. For another development in the "Challenger" biplane series, the model C-3 with Warner "Scarab" engine, see the discussion for A.T.C. #97 in this volume.

Listed below are specifications and performance data for the "Comet 130" powered "Challenger" biplane model C-4A (KR-34A); span upper 30'1", span lower 29'2", chord both 62", wing area 285 sq. ft., airfoil Aeromarine 2A Mod., length 23'2", height 9'3", empty wt. 1331, useful load 1069, payload

Fig. 277. Kreider-Reisner "Challenger" with 4 cylinder "Caminez" engine, flown by Amos Kreider in 1928 National Air Tour.

500, gross wt. 2400 lb., max. speed 115, cruise 98, land 45, climb 615, ceiling 14,500 ft., gas cap. 50 gal., oil 4 gal., range 550 miles. Price at the factory averaged $6500. The fuselage framework was built up of welded chrome-moly steel tubing, faired to shape with wood fairing strips and fabric covered. The wing panels were built up of solid spruce spars that were routed to an "I beam" section and spruce and plywood built-up ribs, also fabric covered. The fabric covered tail-group was built up of welded chrome-moly steel tubing, the fin was ground adjustable and the horizontal stabilizer was adjustable in flight. The cockpits were roomy and deep and an easy entrance door was provided for the passengers, baggage was carried in a compartment behind the pilot's cockpit. Standard equipment included a metal propeller, wheel brakes, and an inertia-type engine starter. The C-4A was one of the first airplanes to use the Westinghouse "Micarta" propeller, the blades were formed from a resin-impregnated fabric. "Micarta" was a material of very desirable qualities, but not particularly satisfactory for propeller blades. The next development in the "Challenger" C-4 series was the C-4B (KR-34B) that was powered with the 5 cylinder Wright "Whirlwind" J6 engine of 165 h.p.

Perhaps it would be interesting to some to know some of the mysteries of aircraft designation; here is a short explanation of the Kreider-Reisner designation system. C-1 was a designation used for "commercial...first model"; C-3 was "commercial...third model"; C-4 was "commercial...fourth model", and so on. Later, this was changed to a different system following this example: KR-31 would be Kreider-Reisner-"3 place...first model"; KR-34 would be Kreider-Reisner-"3 place... fourth model", and so on. The suffix letter showed modification of that particular type, usually this change was only effected by a different engine installation.

A.T.C. #89
(11-28)
FAIRCHILD, "MODEL 71"

Fig. 278. The stately "Fairchild" model 71 was powered with a "Wasp" engine and carried 7 passengers and baggage.

The stately Fairchild model "71" as pictured here in the various views, was a modification of the earlier model FC-2W2, and was a culmination of all the features and requirements that were dictated by many years of service experience. An experience accumulated on the rugged "frontiers" of aviation in the many different parts of the world where "Fairchild" monoplanes had been serving. The "Model 71", in it's time, became quite a "work-horse" of the airways, and was by far the most popular of the "Fairchild" cabin monoplanes for many years.

Although substantially improved, it was still fairly typical of the earlier models in the "FC" series, still bearing the familiar strut-braced thick sectioned semi-cantilever wing that could also be folded back easily when and if necessary. In outward appearance the "71" was cleaned up and faired out much better here and there, and the cabin section had quite a large carrying capacity (145 cu.ft.) for passengers and all manner of cargo. True to established Fairchild tradition, the per-

Fig. 279. The "71" was often called the champion of the bush-country, shown here on Fairchild built floats.

Fig. 280. Fairchild "71" on skis, serving as air-borne freighter in Canada.

formance was very good for a ship of this size and it's ability to get in and out of small air-fields with a sizeable load, whether on wheels, skis, or floats, made it especially popular in the so-called "bush country" of the U.S., Alaska, and Canada. It also served long and well in Mexico, and a number of South American countries.

In this country it was having a good measure of acceptance, and proved ideal and a money-maker too, on numerous feeder-line air routes that fed the transcontinental airways system. In so many words, the "Seventy One" was an extremely versatile airplane that was docile and dependable, and was daily used in many types of flying in a very many different places. It was often remarked that the "Fairchild 71" was a "motherly" looking airplane that was just as gentle and trustworthy as it's appearance portrayed. Many have called the "71" the "champion of the bush", and a good number were still flying up to and including the fifties". As stories go, it's been said that there are absolutely no "Seventy-Ones" available second-hand, unless you might want to pay as much as twice their original price!

In it's basic form, the "Model 71" was a high wing cabin monoplane that was powered with a 9 cylinder Pratt & Whitney "Wasp" engine of 420 h.p., it had more than ample seating for 7 and carried a useful load of some 2500 lbs., of which more than 1400 lbs. was allowable for payload. The type certificate number for the "71" was issued in Nov. of 1928; this model remained quite popular and was continuously built on thru 1930. It went into a modification called the "71A" and then was finally replaced by the "Fairchild 100", which later became the

Fig. 281. Well suited to the requirements of business, this "71" was in service for Dahlberg Sugar Co.

"Pilgrim 100". For details of the "Model 71A", see discussion for A.T.C. #289; for details of the "Model 100", see discussion for A.T.C. #390. In 1929 and thru 1931, the "Seventy-One" was also built in various military versions as a photo-plane or as a utility cargo-carrier, doing well in either capacity.

Listed below are specifications and performance data for the "Wasp" powered Fairchild "Model 71"; wing span 50'2", chord 84", wing area 332 sq. ft., airfoil Gottingen Mod., length 33' height 9'4", the following weights are for the earliest version, empty wt. 2700, useful load 2500, payload 1427, gross wt. 5200 lb., max. speed 138, cruise 110, land 52, climb (420 h.p.) 980, ceiling 15,500 ft., gas cap. 148 gal., oil 12 gal., range 650+ miles. Most of the Model "71" however were as follows; (landplane), empty wt. 2930, useful load 2570, payload 1500, gross wt. 5500 lb., max. speed 135, cruise 110, land 57, climb (420 h.p.) 900, ceiling 15,300 ft., gas cap. 148 gal., oil 12 gal., range 650+ miles. As a seaplane with "Fairchild" pontoons; empty wt. 3270, useful load 2230, payload 1160, gross wt. 5500 lbs., max. speed 129, cruise 105, land 57, climb 875, ceiling 15,000 ft., range 630 miles. Price at the factory as landplane was $18,900., price as seaplane was $22,400. Standard equipment included metal prop, wheel brakes, inertia-type engine starter, and cabin heater. The latest and heaviest version weighed as follows; empty wt. 3130, useful load 2370, payload 1240, gross wt. 5500 lb., max. speed 138, cruise 110, land 57, climb (450 h.p.) 1000, ceiling 16,000 ft. gas cap. 160 gal., oil 15 gal., range 770 miles. The fuselage framework was built up of welded chrome-moly steel tubing, faired to shape and fabric covered. The wing framework was built up of spruce box-type spars and spruce and plywood truss-type ribs, also fabric covered. Fuel tanks were mounted in the wing. The fabric covered tail-group was built up of welded chrome-moly steel tubing, the fin was ground adjustable and the horizontal stabilizer was adjustable in flight. The baggage area was in the cabin aft.

Fig. 282. U. S. Army Air Corps YF-1, a photo-ship which was basically a Fairchild "71".

A.T.C. #90
(11-28)
LOENING "AIR YACHT", C2C

Fig. 283. "Cyclone" powered Loening C2C was used by T.A.C. on run from Detroit to Cleveland, 100 mile trip was made in 75 minutes from dock to dock.

This model was the improved version and the latest modification of the Loening "Air Yacht" series; information gained through many hours of service experience in many different parts of the country had brought forth the necessity for numerous but minor changes. The Loening "Air Yacht" model C2C was powered with the latest and more powerful version of the Wright "Cyclone" engine that was rated at 525 h.p. This engine also had many improvements and it's most noticeable addition was the exhaust collector-ring that was mounted in front of the cylinders; a similiar method was later used on the new J6 series "Whirlwind". Normal seating of the model C2C as a transport was for 7 places, but Loening also offered this ship in custom built versions with specially arranged seating and plush interiors as a "sportsman's model", or for fast functional luxury travel for the busy executive. Justifiably so, Loening was very proud of the fact that the roster of "Air Yacht" owners read like a "who's who" in the world of sports, journalism, finance, big business, and in the aviation industry itself.

As shown here, the "Air Yacht" amphibian was proving it's worth and versatility on many short-haul routes connecting large cities bordering on lakes and rivers; offering rapid and frequent service from "downtown to downtown", this was only possible through the capabilities of the amphibious airplane. Remembering so vividly, it gives pause to wonder how many others can recall the very distinctive exhaust tone and propeller noise of the "Cyclone" powered Loening "Air Yacht", especially on take-off and when grabbing for altitude. They had a terrific vibrating "tone" that was easily heard for some miles around, and would but surely put to shame even the later day "BT-13" and "AT-6" which as many will remember were also very well noted for their terrific "tone". The "Air Yacht" shown here, was part of a fleet operated by "T.A.C." (Thompson Aeronautical Corp.) on a run from Detroit to Cleveland. Running three trips daily, flying the 100 mile trip in 75 minutes from dock to dock; this service was from "downtown to downtown". Other ferry-services were soon established in Seattle, Wash. and in Oakland, Calif.

Listed below are specifications and performance data for the "Cyclone" powered

Fig. 284. Loening "Air Yacht" C2C, with Wright "Cyclone" engine.

Loening "Air Yacht" model C2C; span upper and lower 46'8", chord both 72", wing area 517 sq. ft., airfoil "Loening", length 34'8", height wheels down 13'2", height wheels up 11'5", empty wt. 3894, useful load 2006, payload 996, gross wt. 5900 lb., max. speed 124, cruise 102, land 52, climb 870, ceiling 13,800 ft., gas cap. 140 gal., oil 10 gal., range 550 miles. Price at the factory for standard model was $27,900. The following figures were for a later version; empty wt. 3894, useful load 2356, payload 1346, gross wt. 6250 lb., performance of this version suffered slightly in some instances. Construction details of the model C2C were typical, refer to previous discussions. Cabin interiors and arrangements were made to order, running the gamut from a plush 4 place model with lounges, etc., to a coach-type model seating 8 passengers and a pilot; these were used on air-ferry service involving only short hops. Previous discussions on the Loening "Amphibian" were A.T.C. #34, #66, and #67, all in this volume. The next development in the "Air Yacht" series was the C2H, see discussion for A.T.C. #91 in this volume.

Fig. 285. Chas. A. Lindbergh and his private "Air Yacht".

A.T.C. #91
(11-28)
LOENING "AIR YACHT", C2H

Fig. 286. A "Hornet" powered C2H on factory ramp, being readied for delivery.

The Loening "Air Yacht" C2H version was just about identical to the model C2C in most all respects except for the powerplant installation, which in this case was the new improved 9 cylinder Pratt & Whitney "Hornet" engine of 525 h.p. Pictured here in one of the views is a "Hornet" powered C2H that had just landed in the bay and is taxiing toward the dock, where the wheels will be let down and the airplane will run right up the ramp to discharge it's load of passengers high and dry. This versatile and sensible type of service from "downtown to downtown" was catching the public's fancy and was indeed only possible with an amphibious aircraft.

These unusual airplanes by their very nature had prompted the forming of "air-ferry" service in various parts of the country; one such air-ferry from San Francisco to Oakland, was giving six minute service from dock to dock, all day long. Another such service plied the short stretch from Seattle to Bremerton (15 miles), giving quick straight-line service which saved commuters a lot of time. Other services were started here and there, offering scheduled and charter flights that could take one from point to point regardless of whether it involved a landing on water or an inland air-strip, the Loening "Air Yacht" took it all in stride.

The type certificate number for the "Hornet" powered C2H was issued in November of 1928, and it was about this time that "Keystone Aircraft" (formerly Huff-Daland) of "Keystone Bomber" fame was merged together with "Loening Aeronautical" to make it Keystone-Loening. The factory where the Loening "Air Yachts" were built was in New York City with a spacious dock adjacent to the factory and a on-and-off ramp leading into the East River.

Listed below are specifications and performance data for the "Hornet" powered Loening "Air Yacht" model C2H; span upper and lower 46'8", chord both 72", wing area 517 sq. ft., airfoil "Loening", length 34'8", height wheels down 13'2", height wheels up 11'5", empty wt. 3894, useful load 2356, payload 1346, gross wt. 6250 lb., max. speed 124, cruise 102, land 55, climb 820, ceiling 13,000 ft., gas cap. 140 gal., oil 10 gal., range 550 miles. Price at the factory for the standard model was $27,900. Cabin interiors and arrangement were made to order, running the gamut from a deluxe 4 place "air yacht"

Fig. 287. "Hornet" powered C2H taxiing towards dock, plane will run up ramp and discharge passengers high and dry.

with lounges and many conveniences, to a coach model seating 8 passengers and a pilot; these 9 place versions were used on air-ferry service involving only short hops. The fuselage and float framework was built up of wood members that were gusseted with metal at every joint, a protective coating was applied then the framework was covered with "Alclad" sheet. The wing framework was built up of laminated spruce spars and aluminum alloy stamped-out ribs, the wing framework was fabric covered. The fuel tank was mounted in the lower fuselage under the pilot, the oil tank was in the engine nacelle. A 3-bladed metal propeller, wheel brakes, and an electric inertia-type engine starter were standard equipment. The wings were wired for lights. The next development in the "Loening" amphibian series was the 4 place "Commuter", which was built under A.T.C. #219.

Fig. 288. This Loening C2H was used to transport executives, and for oil field exploration.

A.T.C. #92
(12-28)
PITCAIRN "SUPER MAILWING", PA-6

Fig. 289. The Pitcairn "Super Mailwing" model PA-6 was powered with a Wright J5 engine and could carry over 500 lbs of payload.

In company with the hard-working "Stearman" C3B, the Pitcairn "Mailwing" series were becoming the most popular airplanes on the various short haul feeder-line mail routes around the breadth and length of the country. By their very nature and the work they performed, they were somewhat symbolic of a sort of modern day "Pony Express".

The Pitcairn "Super Mailwing" as shown here, was a development from the earlier "Mailwing", the model PA-5. After a good many hours and thousands of miles of service experience with the earlier version, Pitcairn decided upon a few apparent refinements and modifications that were needed to meet the requirements imposed by the greater demands for airplanes of this type. These changes and improvements were all incorporated into the new PA-6 "Super Mailwing" design. Still, the new "Super" was basically similiar to the earlier PA-5 with the exception of an enlarged cargo hold that now had a capacity of 40 cu. ft., and there was a slight boost in the allowable payload; this without any appreciable loss of performance. The fuselage cross-section was deepened and lengthened with some improvements made in the pilot's cockpit, and a rounding off of the nose-section; the familiar "spinner" of the first few "Mailwings" was now gone. The landing gear was modified to some extent, mostly by the location of attachment, and the "oleo legs" were faired with streamlined metal "cuffs". There was a slight change in the center-section strut layout, but otherwise the wing cellule remained the same.

Also powered with a 9 cylinder Wright "Whirlwind" J5 engine of 220 h.p., the PA-6 had a flashing performance, but was not quite as lively as the earlier PA-5; although in the end it was proved better suited for the conditions prevailing on the "routes" at this time, carrying on and doing it's part in the best "Mailwing" tradition. Typical in the "mail version", the model PA-6 was a single place open cockpit biplane with a hatch-covered compartment up forward holding over 40 cu. ft. of air-mail and cargo. The "Super

Mailwing" was also available in a sport version for the sportsman-pilot that was a 3 place open cockpit type and was called the "Super Sport Mailwing", performance of the two models was more or less comparable. Typical of every "Pitcairn" airplane ever built, clearback to the very first, the "Super" and the "Sport" were trim, beautiful, and well designed airplanes of sure-footed and flashing performance, with a record of dependability that was well known throughout the industry. The model PA-6 was first under a Grp. 2 approval numbered 2-22, but this approval was superceded by A.T.C. #92 which was issued in December of 1928. The certificate was revised and re-issued in April of 1929 to include the 3 place "Super Sport Mailwing".

Listed below are specifications and performance data for the "Whirlwind" J5 powered Pitcairn "Super Mailwing" model PA-6; span upper 33', span lower 30', chord upper 54", chord lower 48", wing area 252 sq. ft., airfoil "Pitcairn", dihedral upper 1 deg., lower 4 deg., length 23'4", height 9'2", empty wt. 1755, useful load 1295, payload 695, gross wt. 3050 lb., max. speed 128, cruise 109, land 52, climb 900, ceiling 16,000 ft., gas cap. 70 gal., oil 10 gal., range 600 miles. Price at the factory was $8500 to $11,500, according to the equipment desired. The following figures are for a later version; empty wt. 1892, useful load 1158, payload 558, gross wt. 3050 lbs., performance in this version was comparable. The "Sport" version was usually operated with a lesser gross weight, so performance was bettered in all instances. The fuselage framework was built up of welded chrome-moly steel tubing in a combination of both round and square section, heavily faired to shape with wood and metal fairing strips and fabric covered. The wing panels were built up of solid spruce spars and spruce and plywood built-up ribs, also fabric covered; ailerons were on the lower panels only. The landing gear was of the out-rigger type and the "oleo struts" were now covered with streamlined metal "cuffs". The fabric covered tail-group was built up of welded chrome-moly steel tubing, the fin was ground adjustable and the horizontal stabilizer was adjustable in flight. A metal propeller, wheel brakes, and a hand crank inertia-type engine starter was standard equipment; the "mail version" was completely equipped for night-flying. The next development in the "Mailwing" series was the PA-6B and PA-7, which were built under A.T.C. #196.

Fig. 290. This is the "Super Sport Mailwing" version of the model PA-6, note covered passenger's cockpit that normally could carry two.

A.T.C. #93
(12-28)
LOCKHEED "WASP-VEGA 5"

Fig. 291. The clean and uncluttered design of the "Vega" was pure symmetry; performance with the P & W "Wasp" engine was sensational.

Being a logical development of the earlier "Whirlwind-Vega", the "Wasp-Vega 5" was also a 5 place high wing cabin monoplane of typical Lockheed configuration and construction. That is, it had an all-wood monococque fuselage and an all-wood full cantilever wing; the "Vega 5" now mounted the more powerful 9 cylinder Pratt & Whitney "Wasp" engine of 400-425 h.p. In the quest for still higher performance, the "Wasp" engine was installed in the standard "Vega"; the performance derived from this added power was a combination that was nearly sensational. It was only natural then for these "Wasp" powered "Vegas" to go right out and start breaking records right and left!

The famous "Yankee Doodle" shown here, was the very first "Wasp-Vega" of this type and was flown from Los Angeles to New York by Art Goebel with Harry Tucker aboard, when they broke the transcontinental record that was formerly held by Macready and Kelly. These two were Army Air Service pilots that spanned the continent in a "Fokker" T-3 in May of 1923, their time was just under 27 hours. Goebel and Tucker's time was just under 19 hours. A return trip record from New York to Los Angeles of just under 25 hours was set by C. B. D. Collyer and Harry Tucker in the same "Yankee Doodle" later that year in Oct. of 1928. The Los Angeles to New York record was then broken by Frank Hawks in a "Wasp" powered Lockheed "Air Express" in Feb. of 1929, setting a time of just under 18 hours. Hawks also shattered the N.Y. to L.A. record in a time of just under 19 hours, and so the race was on! From then on it was "Lockheed" beating another "Lockheed", for some time to come. Graphically proving the company's slogan that "It takes a Lockheed to beat a Lockheed".

Wiley Post's first "Winnie Mae", a sistership to the later and more famous "Winnie Mae" that set a round the world record or two, was also a "Vega 5" of this type and was built late in 1928; it was about the 4th "Wasp-Vega" to come off the line. A "Wasp-Vega 5" fitted with "floats", trail-blazed a route from Seattle to Alaska in April of 1929 for a proposed air-line that was to be operated by the Alaska-Washington Airways. "Lockheed" accomplishments followed one another in rapid succession and it was probably the most talked-about airplane in the industry during this period of time! The type certificate number for the "Wasp-Vega 5" was issued in Dec. of 1928 and then re-issued in Sept. of 1929 as a float-seaplane powered with a

Fig. 292. Famous "Yankee Doodle" was a 1928 "Wasp-Vega", set numerous cross-country records.

"Wasp" engine of 450 h.p.; allowing a total gross load of 4698 lbs. The "Wasp-Vega" was built by the Lockheed Aircraft Corp. at Burbank, Calif. "Jack" Northrop had left by now to make intensive studies in all-metal construction and flying-wing design, but before he left he laid out the engineering for both the "Wasp-Vega" and the "Air Express" models. "Gerry" Vultee had taken over as chief engineer and Lee Schoenhair was the chief test pilot. Lockheed "stars" were shining bright, the outlook was great, and the plant was already employing some 50 people.

Listed below are specifications and performance data for the "Wasp" powered Lockheed "Vega 5"; wing span 41', chord at root 102", chord at tip 63", M.A.C. 80", wing area 275 sq. ft., airfoil at root Clark Y-18, airfoil at tip Clark Y-9.5, length 27'8", height 8'6", empty wt. 2361, useful load 1672, payload 1012, gross wt. 4033 lb., max. speed 170, cruise 140, land 54, climb 1300, ceiling 20,000 ft., gas cap. 96 gal., oil 10 gal., range 725 miles. Price at the factory in July of 1928 was listed at $18,500. The

Fig. 293. The "Wasp-Vega" maintained fast schedules on a number of early air-lines.

Fig. 294. "Yankee Doodle" shown here at the 1928 Air Races held in Los Angeles.

following figures are for a later version; empty wt. 2492, useful load 1541, payload 790, gross wt. 4033 lb., performance figures were not affected. The fuselage framework was built up of two laminated plywood "shells" that were formed to shape in a mold called a "concrete tub", and then assembled over circular shaped wood formers that were held in line by a few wood stringers; after all cut-outs were made the fuselage was fabric covered for increased strength and better finish. The wing framework was built up of solid spruce spars and spruce and plywood built-up ribs, after assembly the wing framework was covered with a plywood veneer and also fabric covered for increased strength and better finish. The tail-group was all wood and of similiar construction. The collector-ring for the engine's exhaust gases was submerged in the cowling, and the rear cowling was profusely louvered for cooling and ventilation. Fuel tanks were mounted in the wing, one either side of the fuselage. A metal propeller, wheel brakes, and an inertia-type engine starter were standard equipment. For the next Lockheed development see discussion for A.T.C. #102 which discusses the "Air Express'.

A.T.C. #94
(12-28)
HAMILTON "METALPLANE", H-47

Fig. 295. Hamilton "Metalplane" model H-47, powered with P & W "Hornet" A-2 engine of 525 h.p.

Pressured by events, the next development in the Hamilton "Metalplane" series was the improved model H-47. Designed to keep pace with changing demands, with a desire to offer higher cruising speeds and a better performance in the interests of better service. The "Metalplane" in the model H-47 had been slightly modified over the earlier model H-45, but was still very much typical in most all respects. In it's basic form as pictured here in the various views, the H-47 was also an 8 place high wing cabin monoplane with a thick full cantilever wing; the wing, the fuselage, and the tail-group were also of all-metal riveted construction using a corrugated "Alclad" metal skin covering throughout. The powerplant for this new version was the larger and more powerful Pratt & Whitney "Hornet" engine of 500-525 h.p.; the added power giving this ship a much greater utility and a nice boost in all-round performance. The performance race amongst the air-lines was already in evidence, in an effort to gain a competitive advantage. Several airlines of this period, such as "Northwest" and "Universal", to name a few, used the "Hamilton" with great success on their scheduled daily runs.

The model H-47 was first built under a Grp. 2 approval numbered 2-14 which was issued in November of 1928. This approval was then superceded by A.T.C. #94 which

Fig. 296. Hamilton H-47 in service with Northwest Airlines.

Fig. 297. The Hamilton H-47 carried 7 passengers at a cruising speed of 121 m.p.h.

was issued in December of 1928; and re-issued later for added wing area, a higher allowable gross weight, and some further minor modifications in April of 1929. The model H-47 was also built under a Grp. 2 approval numbered 2-125 as a 7 place twin-float seaplane; another version was built under a Grp. 2 approval numbered 2-129, this version was powered with a 525 h.p. Wright "Cyclone" engine and was an extremely rare type in this series. The "Metalplane" was manufactured by the Hamilton Metalplane Co. at Milwaukee, Wis., a division of the Boeing Airplane Co. at Seattle, Wash.

Listed below are specifications and performance data for the "Hornet" powered Hamilton "Metalplane" model H-47; wing span 54'5", chord at root 108", chord at tip 75", wing area 387 sq. ft., airfoil Clark Y Mod., length 34'8", height 9'4", empty wt. 3450, useful load 2300, payload 1290, gross wt. 5750 lb., max. speed 145, cruise 121, land 52, climb 900, ceiling 15,000 ft., gas cap. 140 gal., oil 10 gal., range 600 miles. Price at the factory varied from $24,500 to $26,000. The following are figures of the later version as approved 4-29; wing span 60'5", chord was same, wing area 420 sq. ft., length and height were the same, empty wt. 3699, useful load 2166, payload 1166, gross wt. 5865 lb., performance was not greatly affected. The fuselage framework was built up of riveted aluminum alloy sections that were covered with a corrugated "Alclad" metal skin. The wing framework and the tail-group were built up of similiar construction, the horizontal stabilizer was adjustable in flight. Ailerons were of the off-set hinge type giving lighter and more effective control. The landing gear was typical and the previously used tail-skid had been replaced with a swiveling tail wheel. Interior appointments were improved, and a small rest-room was provided in some instances. Baggage capacity and baggage allowance were increased. A metal propeller, wheel brakes, and an inertia-type engine starter were standard equipment. Complete night-flying equipment was available.

Fig. 298. The Hamilton H-47 was of all-metal construction, note corrugated metal skin covering.

A.T.C. #95
(12-28)
MOHAWK "PINTO" - MLV

Fig. 299. The saucy little Mohawk "Pinto" MLV offered sprightly performance with the "Velie" engine.

The Mohawk "Pinto" series was a line of delightful and refreshing little monoplanes of somewhat controversial characteristics and habits, but more than one have since stated emphatically that when thoroughly mastered, the "Pintos" were a pure joy to fly! The diminutive and saucy looking "Pintos" were low-winged full cantilever monoplanes; a configuration which was still somewhat of an innovation and a brave and lonely approach in the light plane field, in this country at least. Light monoplanes were still not to be trusted too much, and were looked at with a skeptic eye by the average "fly guy" of these times, and especially so if they were low-winged monoplanes. These were approached warily! It is true, that the first "Pinto" type did have a few tricky and nervous habits; one such, was a difficulty to recover every time when wound up tight in an intentional "spin"; two test-pilots including the noted Charles "Speed" Holman had to "bail out" from early "Pintos" for this reason. But this was a fault that was also still somewhat inherent in a number of other airplane types. This nasty and irritating trait was entirely overcome in the later production versions of the "Pinto" series.

The Mohawk Aero Corp. was formed early in 1927 by four young hopefuls not entirely bound by convention, and with an eye to the future; the first "Pinto" built was ready to fly later in that year. Designed by Wallace C. "Chet" Cummings, it was a stubby low-winged full cantilever monoplane seating two side by side, with the seats slightly staggered fore and aft to provide enough shoulder room in the narrow little cockpit; it was comfortable enough but quite chummy. The overall design was very clean and uncluttered, and delivered quite a lively performance with a 60 h.p. Detroit "Air Cat" engine; the 6 cylinder "Anzani" was also used. Many noted pilots, including Chas. A. Lindbergh, flew this original type "Pinto" on numerous trials and shake-down flights; comments were mixed and varied, of course.

The first modification of this original configuration also seated two, but now in individual open cockpits placed in tandem, and it was powered with the new 7 cylinder Warner "Scarab" engine of 110 h.p.; this was a spirited little wild-cat with a terrific performance. This same airplane was later flown in the 1928 National Air Tour by the amiable Dr. Jos. A. Nowicki, who loved it dearly and praised it highly. It's been said that flying this "Pinto" version was somewhat like having a tiger by the tail; but it was sheer fun! Another modification was to be set up for a production model, it was quite similiar to the Warner powered "Pinto" as described

Fig. 300. Mohawk "Pinto" in Canada.

above, except that it was powered with the new 5 cylinder "Velie" engine of 55 h.p. This model was the "MLV" as shown here in one of the views, it was the first of the "Pinto" series to be certificated; it's type certificate number was issued in December of 1928.

The MLV and all the previous models, had the "oleo leg" of the landing gear mounted on the outside of the wheel hub providing a wider stance for better ground stability but this was a very inconvenient feature in time of wheel removal and landing gear repair; this was consequently remedied on the later models by putting the "oleo leg" on the inside of the wheel hub in a more conventional manner. Not yet entirely satisfied with it's performance characteristics, nor with the reception the "Pinto" was getting up to now, Mohawk had their designs modified somewhat by Prof. John D. Akerman and brought these out early in 1929 as the "Spurwing" and the "Redskin". These were both Warner "Scarab" powered; the "Spurwing" was a 2 place with open cockpits in tandem and the "Redskin" was a 2-3 place type with an enclosed canopy. Labeled as prototypes, further development of these two models led to the M1C-W and M1C-K.

Starting out early in 1927 as the Mohawk Aero Corp. of Minneapolis, Minn., the firm was reorganized later in the year as the Mohawk Aircraft Corp. with Leon A. Dahlem as president and Sumner Whitney as treasurer, Wallace C. "Chet" Cummings was in charge of design and flight tests were under the supervision of George A. MacDonald.

Listed below are specifications and performance data for the "Velie" powered Mohawk "Pinto" model MLV; wing span 30'6", chord at root 66", chord at tip 51", wing area 124 sq. ft., airfoil U.S.A. 35 Mod., length 20'2", height 6'3", empty wt. 858, useful load 474, payload 177 gross wt. 1332 lb., max. speed 102, cruise 89, land 38, climb 570, ceiling 8,000 ft., gas cap. 22 gal., oil 2 gal., range 400 miles. Price at the factory in 1928 was $2875. The prototype version of the "Pinto" was a good deal lighter and performance was considerably better, but this phenomonon happens to all airplane types; the final production version is most always heavier with a proportionate loss of performance. The fuselage framework of the "Pinto" was built up of welded chrome-moly steel tubing, heavily faired to shape with wood fairing strips and fabric covered. A fuel tank of 22 gallons was placed in the upper fuselage ahead of the forward cockpit. The full cantilever wing framework was built up of spruce and mahagony plywood spars of box-type section, the ribs were built up of mahagony plywood webs and spruce cap-strips, the leading edge up to the front spar was covered with mahagony plywood to add strength and to preserve airfoil shape, after assembly the wings were fabric covered. Oddly enough, the wing was in two panels; either panel could be removed by the removal of 4 large bolts. Ailerons were of the projecting hinge type and were torque tube operated, rudder and elevators were operated by cables that were completely enclosed in the fuselage, no "horns" protruding into the slip-steram. The fabric covered tail-group was built up of welded chrome-moly steel tubing, the fin was ground adjustable and the horizontal stabilizer was adjustable in flight. The robust landing gear was a wide tread split-axle type, upper ends of the "oleo legs" were attached to the front spar, wheel tread was 6 feet. The tail-skid was of the trouble-free spring leaf type. The colorful "Pinto" color scheme of scallops in contrasting cream and black, or cream and red, was created by Douglas Rolfe, the noted "Air Progress" author who is featured in model airplane magazines of recent times. The model MLV was also available with the 5 cylinder LeBlond engine of 60 h.p.

Fig. 301. Various engines were tested in the "Pinto"; this ship was first tested with Cirrus Mk. III.

A.T.C. #96
(12-28)
FOKKER "TRI-MOTOR", F-10A

Fig. 302. Fokker "Tri-Motor" F-10A over Miami, Pan American Airways had twelve of these on Miami-Havana-West Indies service.

The stately Fokker "tri-motored" F-10A was a modification and a general refinement of the previous model F-10, which was redesigned to incorporate many useful changes that were deemed advisable through actual service experience. Experience that was gained on a number of air-lines across the country. The earlier F-10 certainly was a fine ship, but constant improvement and devotion to minor detail was becoming necessary to keep pace with stricter demands, a clamor imposed by a rapidly increasing interest in air-travel, and an increase in the number of air-carriers desiring this type of equipment. These demands required action which led to the development of the improved model F-10A.

In it's basic form, the tri-motored F-10A was still quite typical in it's unmistakable "Fokker" configuration; a large high wing cabin monoplane with an all-wood full cantilever wing that was powered with 3 Pratt & Whitney "Wasp" engines of 420-450 h.p. each, each engine swung a three-bladed metal propeller. The cabin capacity seated twelve passengers with many added niceties to the interior that offered greater appeal and a commodious comfort. Other changes incorporated in the F-10A, though hardly noticeable, offered greater convenience for faster servicing, and a slight increase in all performance requirements. This new F-10A version was in a sense a haughty looking airplane with a peculiar carriage that seemed to imply a great pride of lineage, and very well it might, because "Fokker" airplanes by now had a background of achievements well worthy of notice and praise; in this case a little apparent snobbishness could usually

Fig. 303. Poised for take-off, this Fokker F-10A was in service with Transamerican Airlines Corp.

Fig. 304. The "Townend ring" anti-drag engine cowling boosted top speed of F-10A by 8 m.p.h. "Townend ring" straightened out airflow around engine cylinders and reduced drag.

be overlooked. The "Fokker" type were sturdy and dependable, they seemed to mellow with age, and there were still a good number of the F-10A in active daily use as late as 1935.

The type certificate number for the "Fokker" tri-motored F-10A was issued in December of 1928 for 3 "Wasp" engines of 420 h.p. each and a gross load of 12,500 lb., and was then re-issued in April of 1929 for the installation of three "Wasp" engines of 450 h.p. each and an allowable gross load of 13,100 lb. Later in 1929 and early 1930, Fokker experimented with the "Townend ring" type of anti-drag engine cowlings for the F-10A, first only on the outboard engines and finally on all three engines. Exacting tests proved the use of these anti-drag cowlings as very worthwhile. They added at least 8 m.p.h. to the maximum speed (153.7) and at least 3 m.p.h. to the most economical cruising speed. All tests were made with a useful load of 5320 lb. and a gross load of 13,100 lb. Propeller settings were then also changed for a much better all-round performance.

Listed below are specifications and performance data for the P & W "Wasp" powered Fokker "Tri-Motor Deluxe" model F-10A; wing span 79'2", wing chord tapered in planform and section, wing area 854 sq. ft., airfoil "Fokker", length 50'7", height 12'9", empty wt. 7780, useful load 5320, payload 2800, gross wt. 13,100 lb., max. speed 145, cruise 123, land 60, climb 1250, ceiling 18,000 ft., gas cap. 360 gal., oil cap. 30 gal., range 6½ hours or 765 miles. Price at the factory averaged $67,500. The following figures are for one of the last in this series; empty wt. 8500, useful load 5500, payload up to 3000 lb., gross wt. 14,000 lb., performance in this version suffered only slightly. The fuselage framework of the F-10A was built up of welded chrome-moly steel tubing, faired to shape and fabric covered. The huge one-piece full cantilever wing panel was built up of laminated spruce spars, spar flanges, and plywood web ribs; reinforced with plywood stringers and covered completely with plywood veneer. The fabric covered tail-group was built up of welded chrome-moly steel tubing, the fin was ground adjustable and the horizontal stabilizer was adjustable in flight, rudder and elevators used "aerodynamic balance" to ease control pressures. The ailerons were of the off-set hinge, or "Freise" type, for more effective control. The landing gear was shockcord sprung on the first few of the Fokker "Tri-Motors", which is rather unusual for a ship of this size; all later models had "oleo" shock struts. Three-bladed metal propellers were standard equipment on the F-10A, as were inertia-type engine starters, and wheel brakes. Earlier models still had the familiar "tail-skid", but this was changed to a "tail-wheel" on the later models. The earlier models acquired all modifications from the factory, as the modifications were released, so consequently the older models were being kept up to date even though in service. For the next "Fokker" development see discussion of J6-300 powered "Standard Universal".

Fig. 305. Swinging the compass on an "F-10A" at the factory field in Teterboro Field, N. J.

A.T.C. #97
(12-28)
KREIDER-REISNER "CHALLENGER", C-3

Fig. 306. Kreider-Reisner "Challenger" biplane, model C-2 type. Powered with Warner "Scarab" of 110 h.p. Tested as prototype for model C-3.

The Kreider-Reisner C-3 was also a typical open cockpit "Challenger" biplane and was the third model in this series; it was a neat and good looking airplane that proved to be a fairly good engine-airplane combination. A combination that should have appealed strongly to the small operator or for personal use by the private-owner. In spite of it's attractive qualifications and promising possibilities, for some reason it was destined to remain an extremely rare type. In it's basic form, the model C-3 was a 3 place open cockpit biplane and was comparable to the other "Challenger" models. Comparable in most all respects except for the engine installation which in this case was the spunky and increasingly popular 7 cylinder Warner "Scarab" engine of 110 h.p.

The type certificate number for the Kreider-Reisner model C-3 was issued in December of 1928 and this was the last certificate number to be issued for that year. Of some interest might be the fact that 89 Warner "Scarab" powered airplanes were reported built in 1928, this list would include models such as the Stinson "Junior" SM-2, Cessna AW, Consolidated "Husky Jr.", the "Challenger C-3" mentioned here, and prototype versions of the Mohawk "Pinto", Berliner "Parasol", Buhl "Junior Airsedan", "Travel Air", and of course a good number of experimental prototypes.

December 17th of 1928 was the 25th anniversary of the Wright brothers' first shaky flight at Kitty Hawk, it marked a quarter century of amazing and tremendous development! It is remarkable what had been accomplished in 25 years. The first annual "aircraft show" held at Detroit in the Convention Hall in April 14-21 of 1928, was viewed by at least 150,000 air-minded visitors; 68 airplanes were exhibited and the show was pronounced a great success. Seventy-nine airplanes were later exhibited at the Chicago "aircraft show" in December of 1928; things were beginning to get big for aviation and it was now being looked upon as one of the country's up and coming major industries. For a brief resume of the stature of the industry in 1928, we might mention here that over 3500 commercial airplanes were produced in that year. Of this number some 171 were open cockpit monoplanes, 58 were multi-engined cabin monoplanes, 2348 of this number were open cockpit biplanes thereby proving the popularity of this type; 69 were cabin biplanes, 5 were multi-engined cabin biplanes, 11 were seaplanes and flying boats, and 30 of this number were amphibians. In 1929 this amount of airplanes produced was to nearly double! The approaching year of 1929 was a vertiable riot of activity, it was the most frantic year of all times; it may even seem unbelieveable that 187 "approved type certificates" were issued to 72 manufacturers in that year alone, while there were at least 18 manufacturers experimenting with various models and types!

Getting back to the Kreider-Reisner C-3 that is in discussion here; a "Challenger" biplane of this type was flown by Ted Kenyon to 4th place in the Class A division of the Transcontinental Air Derby from New York to Los Angeles in September of 1928. In this race there were six Warner "Scarab" powered entries; placing a first, second, third, fourth, and eleventh positions. One entry did not finish. Three improved model C-5, which were a later development and a modification of the C-3 type discussed here, were also powered with the 7 cylinder "Scarab" engine. The C-5 were built under a Grp. 2 certificate of approval numbered 2-44, this approval was issued in Feb. of 1929. The "Challenger" biplanes, now available in 3 models, were manufactured by the Kreider-Reisner Aircraft Co. at Hagerstown, Maryland. More than 100 "Challenger" biplanes had been produced up to this time.

Listed below are specifications and performance data for the Warner "Scarab" powered Kreider-Reisner "Challenger" model C-3; span upper 30'1", span lower 29'2", chord both 60", wing area 285 sq. ft., airfoil "Aeromarine Mod.", length 23'6", height 9'3", empty wt. 1165, useful load 835, payload 395, gross wt. 2000 lb., max. speed 110, cruise 95, land 40, climb 590, ceiling 13,500 ft., gas cap. 40 gal., oil 3 gal., range 425 miles. Price at the factory was approx. $5000. The fuselage framework was built up of welded chrome-moly steel tubing, faired to shape with wood fairing strips and fabric covered. The wing panels were built up of routed spruce spars and spruce and plywood built-up ribs, also fabric covered. The fabric covered tail-group was built up of welded chrome-moly steel tubing, the fin was ground adjustable and the horizontal stabilizer was adjustable in flight. The landing gear on the C-3 was of the divided axle type. The next Kreider-Reisner development was the KR-34B, built under A.T.C. #162.

A.T.C. #98
(1-29)
BUHL "SENIOR AIRSEDAN", CA-8A

Fig. 307. The Buhl "Senior Airsedan" model CA-8A, powered with Wright "Cyclone" engine of 525 h.p.

The impressive and stately "Senior Airsedan" was the "big job" and the largest of the Buhl "Airsedan" series. In it's basic form, it was a large 8 passenger cabin biplane or it should more aptly be called a "sesquiplane", because of it's very small lower wings. Except for it's much larger size the "Senior Airsedan" was still quite typical of the other "Airsedan" models in it's basic configuration, it's method of construction, and manner of behavior. The company's designation used for the new "Senior" series was CA-8, and it was produced in two standard versions. The model CA-8A in discussion here and shown in the accompanying views, was powered with the newly improved 9 cylinder Wright "Cyclone" engine of 525 h.p.; the performance of this ship was quite good for an airplane of this size, and it's habits and flight characteristics were pleasant and very commendable. These large "Airsedans" were designed by Ettienne Dormoy and developed primarily as a medium-sized passenger transport; it was sincerely hoped that they would prove attractive as a carrier on the numerous air-lines springing up around the country, that would need a ship of this size for the traffic that would be fed from outlying areas to the larger lines. A few of the "Senior Airsedans" were sold, but rather unfortunately this airplane turned out to be an odd size that was too large for some uses and too small for other chores and didn't fare too well in this respect; possibly due to peculiar circumstances that were coming about and were certainly not the fault of the airplane itself.

The Buhl Aircraft Co. of Marysville, Mich. introduced the "Senior" line of "Airsedans" in the later part of 1928, and the type certificate number for the model CA-8A was issued in January of 1929. This was the first certificate issued in that year and before the year was over, some 187 "approved type certificates" were issued by the "Dept." to some 72 manufacturers! This was by far the biggest year in aviation history from the standpoint of the unusually large number of manufacturers engaged in building of the most diversified selection of airplanes ever to come before the flying and buying public. However sad, this phenomonon lasted about a year and was nipped in two by the ravaging effects of the "bank crash" and the "big depression", and has never happened again since then.

Listed below are specifications and performance data for the "Cyclone" powered Buhl "Senior Airsedan" model CA-8A; span upper 48', span lower 31', chord upper (constant) 96", chord lower tapered to 63.5"

M.A.C., wing area 462 sq. ft., length 36', height 10', wheel tread 12', empty wt. 3542, useful load 2658, payload 1390, gross wt. 6200 lb., max. speed 142, cruise 118, land 48, climb 950, ceiling 18,000 ft., gas cap. 181 gal., oil 15 gal., range 775 miles. Price at the factory was $19,500. This model was also available with the newly improved 9 cylinder Pratt & Whitney "Hornet" engine of 525 h.p. as the model CA-8B, see discussion for A.T.C. #99 in this volume. Adhering to the general construction principles of the "Buhl" line, the fuselage framework of the "Senior Airsedan" was built up of welded chrome-moly steel tubing, faired to shape and fabric covered. The wing panels were built up of spruce box-type spars and spruce and plywood built-up ribs, the stub-wing was built up of welded chrome-moly steel tubing and was built integral with the fuselage, all panels were fabric covered. The landing gear was attached to the stub-wing. The fabric covered tail-group was built up of welded chrome-moly steel tubing, the fin was ground adjustable and the horizontal stabilizer was adjustable in flight. Both rudder and elevators used overhanging "balance horns" for ease of control. The ailerons also used the "balance horn" but it was inset from the tip and not of the familiar "overhang" type. The "Senior Airsedan" had a landing gear of a "tracking" type that was later used in a slightly modified form on many famous racing monoplanes, planes such as the Travel Air "Mystery Ship", the "Gee Bee" series, and many others. A huge baggage compartment was aft of the cabin section and had a capacity for 50 cu. ft. of baggage and additional cargo. A metal propeller, inertia-type engine starter, wheel brakes, tail-wheel, dual controls, cabin heat and sound-proofing, were included as standard equipment. The next development in the "Airsedan" series was the model CA-6.

Fig. 308. Ettienne Dormoy the designer of the Buhl "Airsedan" series, probably best known for his "flying bathtub" of 1924'

A.T.C. #99
(1-29)
BUHL "SENIOR AIRSEDAN", CA-8B

Fig. 309. Buhl "Senior Airsedan" model CA-8B. A medium-sized transport that carried 7 passengers and a pilot.

This model of the "Senior Airsedan" shown here in various views, was also a medium-sized transport type airplane, and was almost identical to the model CA-8A except for the engine installation; which in this case was the new improved version of the 9 cylinder Pratt & Whitney "Hornet" engine of 525 h.p. The performance and flight characteristics of both of these models could be considered practically the same. The type certificate number for the "Hornet" powered model CA-8B was issued in January of 1929 and it sold at the factory field for $19,500, a price that was well under the average rate for an airplane of this type; the "Senior Airsedan" was also available in a cheaper version that was fitted with a Pratt & Whitney "Wasp" engine of 450 h.p. and sold for $18,500 at the factory field. The performance of the "Wasp" powered airplane was naturally reduced in proportion, of course. The "Wasp" powered version of the CA-8 was built under Grp. 2 approval numbered 2-46.

All of the Buhl "Airsedan" models being manufactured were now of the novel and quite sensible "sesqui-wing" arrangement, the

Fig. 310. A "Wasp" powered "Senior Airsedan" being prepared for delivery at the Buhl Aircraft plant in Marysville, Michigan.

lower wing panels were quite short in overall span and were tapered in plan-form and section, doubling in duty by acting as a major part of the wing truss, affording a convenient wing-walk, and furnishing an ideal place to fasten the landing gear; all this and they still contributed with some 120 square feet of lifting area. Interior appointments were convenient and tasteful, and cabin walls were insulated and sound-proofed for passenger comfort. Pilot vision was excellent and ground handling was very good for such a large ship. Many features of great appeal were incorporated into these transports, they had much to offer, but they failed to sell in any great number and remained a rather rare type. The next development in the "Airsedan" series was the 6 place model CA-6 which fared much better.

Listed below are specifications and performance data for the "Hornet" powered Buhl "Senior Airsedan" model CA-8B; span upper 48', span lower 31', chord upper (constant) 96", chord lower tapered to 63.5" M.A.C., interplane gap 68", wing area 462 sq. ft., length overall 36', height 10', wheel tread 12', empty wt. 3542, useful load 2658, payload 1390, gross wt. 6200 lb., max. speed 142, cruise 118, land 48, climb 950 ft. first minute, ceiling 18,000 ft., gas cap. 181 gal., oil cap. 15 gal., range 775 miles. The fly-away price at the factory field was $19,500. All construction details were similiar to the model CA-8A, see previous chapter. Like most of the average aircraft manufacturers of this day, Buhl had a small engineering staff that probably never exceeded five men, but Ettienne Dormoy was a brilliant visionary and kept the boys busy with a continuous flow of new ideas and new designs. Those that materialized, were designed, built, and tested, in a matter of months; a far cry from the tedious and lengthy procedures practiced in this modern day and age!

Fig. 311. The "Senior Airsedan" was also offered with the P & W "Wasp" engine with capacity for six.

A.T.C. #100
(1-29)
TRAVEL AIR, MODEL 6000

Fig. 312. The "Travel Air" Model 6000 carried six in comfort, designed to meet the needs of business firms and smaller air lines.

Walter Beech and his enthusiastic crew at "Travel Air" let it be known that they were mighty proud of the model "6000" when it was introduced early in 1928; they were proud and justifiably so, because it was a big beautiful airplane that incorporated the very latest features expressly designed to promote a better acceptance of the convenience and comfort of travel by air. They even went so far as to dub it the "Limousine of the Air" and it was, in a sense, with it's roomy interior and pleasant appointments that were coupled to an air about it that could be best described as luxurious efficiency, whether on the ground or flying.

Pictured here in the various illustrations, we see this "buxom beauty" as a strut braced high wing cabin monoplane of rather large proportions and typical of "Travel Air" construction and basic configuration features, with modifications and improvements added that were later found lacking in the earlier "Travel Air" monoplane, the Model 5000. The new Model 6000 provided ample seating for six, and all were fully enclosed in the cabin section. In many instances previous, the pilot was more or less isolated from his passengers, but it was now being considered as appropriate that to promote a more friendly atmosphere and a spirit of camaradie, it would be better to have the pilot visible and accessible to the whims of the passengers.

The powerplant selected for the Model 6000 was the popular nine cylinder Wright "Whirlwind" J5 engine of 220 h.p., the engine was muffled fairly well and the cabin walls were insulated and sound-proofed to keep noise and vibrations at a fairly low level. Conversing in an airplane aloft at practically a shout, was considered a little old-fashioned by now! The performance of this big beauty was among the best of it's type and it's agility both in the air and on the ground was surprising for a ship of this type and size. They at "Travel Air" were "plugging" this one often and well, and it's various merits were soon recognized; it sold quite well to numerous business-firms and to a number of the newer air-lines around the country. The type certificate number for the Model 6000 was issued in January of 1929 and this model was soon offered with other engine installations listed under their appropriate certificate numbers. The model A-6000-A with "Wasp" engine was built under A.T.C. #116; the model B-6000 with 300 h.p. "Whirlwind Nine" was built under A.T.C. #130.

Although only remotely similiar in appearance, the Model 6000 was actually a development of the earlier 5 place Model 5000 which

Fig. 313. 1928 prototype version of the "Travel Air" Model 6000, proudly announced as the "Limousine of the air".

was a strut braced high wing cabin monoplane of the "Woolaroc" type; an airplane made famous by Art Goebel on his flight across the Pacific Ocean to Hawaii in winning the famous (or infamous, depending on view) "Dole Derby" in August of 1927. Discussion of Grp. 2 approval numbered 2-27 will carry details of this airplane, etc. The Travel Air "Model 5000" is pictured here in one of the views.

The production model of the 6000 that was built under this certificate number, and is shown here, was already improved and modified over the prototype version with features found more suitable to promises extended. Notable among these improvements was an enlarged cabin section for more elbow room, with roll-down windows, removable seats for hauling cargo, and non-shatter safety glass in the pilot's section of the cabin. The tail-skid was replaced with a 14 x 3 steerable tail-wheel, and the rudder shape was modified with a cut-out to suit. The Model 6000 was manufactured by the Travel Air Co. at Wichita, Kan. who was adding new buildings to their plant layout every month or so; it was getting to be the biggest commercial aircraft producer in the country.

Fig. 314. The colorful "Travel Air" Model 5000 was one of the first to offer air-travel in cabin comfort.

Listed below are specifications and performance data for the "Whirlwind" J5 powered "Travel Air" monoplane Model 6000; wing span 48'6", chord 78", wing area 280 sq. ft., airfoil Clark Y-15, length 30'10", height 9'3", empty wt. 2430, useful load 1670, payload 965, gross wt. 4100 lb., max. speed 120, cruise 102, land 55, climb 650, ceiling 12,000 ft., gas cap. 80 gal., oil 6 gal., range 5½ hours or 560 miles. Price at the factory averaged about $12,000. The fuselage framework was built up of welded chrome-moly steel tubing, sound-proofed, insulated, faired to shape and fabric covered. The wing framework was built up of spruce box-type spars and spruce and plywood built-up ribs, also fabric covered. Fuel supply was carried in two wing root tanks of 40 gallons each. This version of the Model 6000 used a "prop spinner", some of the later models did not. The fabric covered tail-group was built up of welded chrome-moly steel tubing, the fin was ground adjustable and the horizontal stabilizer was adjustable in flight. The wide tread landing gear was of the out-rigger type with "oleo" shock-absorber struts. The color scheme was of course optional, but standard colors were black for the fuselage and tail-group, and a bright orange-yellow for the wing; fuselage was striped in chrome-yellow. Metal propeller, inertia-type engine starter, and wheel brakes were standard equipment.

This chapter marks the close of this first volume, a volume depicting the growth of the aircraft industry from a rather shaky and apprehensive start to one that had gained confidence in itself, and began to snow-ball into one of the major industries in this country. Volume Two has us on the threshold of the year 1929, certainly the most memorable and unusual year in the annals of the aircraft industry; it should be interesting and worthwhile to follow this progress.

Fig. 315. Travel Air model "6000" with Wright J5 engine; later converted to model "B-6000" with installation of Wright J6-9-300.

Fig. 316. Line up of "Travel Airs" at factory field; prototype "6000" in foreground.

APPENDIX

BIBLIOGRAPHY

BOOKS:

Aircraft Year Book for 1927-1928-1929-1930
Flying The Arctic by Capt. George Hubert Wilkins
That's My Story by Douglas Corrigan
Air Power For Peace by Eugene Wilson
Around the World in 8 Days by Post & Gatty
"We" by Chas. A. Lindbergh
Spirit of St. Louis by Chas. A. Lindbergh
Conquering The Air by Archibald Williams
Modern Aircraft by Maj. Victor Page
"Flight" by Year
A Chronology of Michigan Aviation by Robert S. Ball
Wings For Life by Ruth Nichols

PERIODICALS:

Western Flying
Aero Digest
Popular Aviation
Alexander "Aircrafter"
National Geographic Magazine
Aviation
The Pilot
Air Transportation
Pegasus, Fairchild Aircraft Div.
Antique Airplane News

SPECIAL MATERIAL:

Licensed Aircraft Register by Aeronautical Chamber of Commerce of America, Inc.

CORRESPONDENCE WITH FOLLOWING FIRMS & INDIVIDUALS:

Alfred V. Verville
Willis C. Brown
Chas. W. Meyers
Grover Loening
Peter Altman
A. W. "Abe" French
Adolf Bechaud
S. J. "Steve" Wittman
Boeing Airplane Co.
Western Air Lines
John W. Underwood
Sikorsky Aircraft Div.
Cessna Aircraft Co.
Spartan Aircraft Co.
Hawaiian Pineapple Co.
Wm. T. Larkins
Eastern Air Lines
Defiance Chamber of Commerce
Lockheed Aircraft Corp.
W. U. Shaw
Pan American World Airways
Fond Du Lac Chamber of Commerce
Temple, Texas Chamber of Commerce
United Airlines
American Airlines
Northwest Orient Airlines
Braniff International Airways
Douglas Aircraft Co.
Delta Airlines
Hartzell Industries

PHOTO CREDIT

Ford Motor Co. from Stephen J. Hudek; Figs. 1 - 41 - 60 - 68 - 70 - 73 - 83 - 84 - 85 - 93 - 94 - 99 - 103 - 115 - 116 - 120 - 123 - 132 - 143 - 162 - 165 - 182 - 228 - 231 - 265 - 267 - 270 - 271 - 272 - 273 - 274 - 275.

Alfred V. Verville; Figs. 2 - 3 - 4 - 6 - 158 - 159.

Buhl Aircraft Co. from W. U. Shaw; Figs. 5 - 48 - 49 - 126 - 308 - 309 - 310.

Boeing Airplane Co.; Figs. 7 - 8 - 9 - 10 - 89 - 90 - 91 - 92 - 108 - 109 - 178 - 214 - 215 - 216 - 217.

Johnson Airplane Co.; Fig. 11.

Smithsonian Institution, National Air Museum; Figs. 12 - 14 - 17 - 21 - 43 - 56 - 67 - 71 - 81 - 87 - 100 - 105 - 111 - 114 - 125 - 135 - 141 - 142 - 163 - 188 - 212 - 213 - 227 - 253 - 256 - 261 - 280 - 294 - 315.

Douglas Aircraft Co.; Figs. 13 - 15 - 16 - 18 - 20 - 22 - 23 - 54 - 55.

Western Air Lines; Figs. 19 - 25 - 184.

B. F. Goodrich Co.; Figs. 24 - 195.

Gordon S. Williams; Figs. 26 - 30 - 31 - 33 - 243 - 276 - 301.

Alexander Film Co.; Figs. 27 - 28 - 29 - 32 - 189 - 190 - 191 - 192 - 193 - 194 - 196 - 197 - 198 - 199.

Institute of Aero Sciences; Figs. 34 - 151 - 170 - 260 - 299.

Marion Havelaar; Fig. 35.

Royal Canadian Air Force; Figs. 36 - 72 - 230.

Ken M. Molson; Figs. 37 - 38 - 88 - 300.

Fokker Aircraft Co.; Figs. 39 - 172 - 183 - 305.

Fairchild Aircraft Div.; Figs. 40 - 42 - 74 - 75 - 76 - 205 - 208 - 277 - 278 - 281.

U. S. Air Corps; Fig. 44.

Stephen J. Hudek; Figs. 45 - 63 - 66 - 69 - 77 - 78 - 96 - 97 - 130 - 186 - 224 - 258 - 283 - 288 - 290 - 306 - 312 - 313.

Chas. W. Meyers; Figs. 46 - 47 - 50 - 51 - 52 - 53 - 104 - 106 - 107 - 144 - 145 - 146 - 155 - 241 - 268.

Roy Oberg; Figs. 58 - 113 - 149 - 150 - 152 - 169 - 179 - 209 - 211 - 284.

Jack McRae; Fig. 59.

Braniff International Airways; Fig. 61.

Reid Studio; Figs. 62 - 160.

Thompson Products Co.; Fig. 64.

Ball Studio; Figs. 65 - 161.

Joseph P. Juptner; Figs. 79 - 80 - 82 - 95 - 147 - 174 - 177 - 207 - 223 - 245 - 252.

Peter M. Bowers; Figs. 86 - 118 - 164 - 176 - 210 - 229.

Northwest Air Lines; Figs. 98 - 263 - 264 - 269.

United Air Lines; Figs. 101 - 110 - 171 - 180.
Macdonald Photo; Fig. 112.
Beech Aircraft Corp.; Figs. 117 - 119 - 138 - 314 - 316.
Travel Air Co.; Figs. 121 - 137.
Phillips Petroleum Co.; Fig. 122
Parks Air College; Fig. 124.
Gerald Balzer; Figs. 127 - 148 - 238 - 279 - 293.
Pratt & Whitney Div.; Figs. 128 - 292 - 295 - 304 - 311.
Grover Loening; Figs. 129 - 221 - 285 - 286 - 287.
U. S. A. F.; Figs. 131 - 187 - 246 - 247 - 282.
Winstead Photo; Figs. 133 - 134.
Pheasant Aircraft Co.; Fig. 136.
Berliner Aircraft Co.; Fig. 139.
North American Aviation Div.; Figs. 140 - 173
Buhl Aircraft Co.; Figs. 153 - 154.
Sanborn Photo; Figs. 156 - 157.
Northrop Corp.; Fig. 166.
Lockheed Aircraft Corp.; Figs. 167 - 168.
American Air Lines; Figs. 175 - 181 - 206 - 289 - 297.
Pan American World Airways; Figs. 185 - 202 - 222 - 302.
Sikorsky Aircraft Div.; Figs. 200 - 201 - 203 - 204.
Cessna Aircraft Co.; Figs. 218 - 219 - 220 - 234 - 236 - 237.
Curtiss Airplane & Motor Co.; Figs. 225 - 226.
Miller Photo from Willis C. Brown; Fig. 232.
Spartan Aircraft Co.; Fig. 233.
Leo J. Kohn; Figs. 235 - 296 - 298.
Fred G. Bell; Figs. 239 - 259.
Willis C. Brown; Fig. 240.
Curtiss Flying Service; Fig. 242.
T. C. Weaver; Fig. 244.
Convair Div.; Figs. 248 - 250 - 251 - 257.
U. S. Navy Photo; Figs. 249 - 254 - 255.
Murdoch Photo; Fig. 262.
Laird Airplane Co.; Fig. 266.
Van Rossem Photo; Fig. 291.
Thompson Aero. Corp.; Fig. 303.
Denkelberg Photo; Fig. 307.